Goodies from the

Yum Yum Tree

*The Internet and Revolution
in the Final Days of Capitalism*

by Alfredo López

Published by Entremundos Publications

461 54th Street

Brooklyn NY 11220

First printing: 2019

ISBN-13: 978-1-877850-02-8

Preface

There's a big bad monster you can't see

who hides behind the Yum Yum tree

He catches little girls and boys

Who reach for the Yum Yum's sweets and toys

But stay together hand in hand

'Cause that's what the monster can't understand

You walk right up and never break apart

And the monster won't know where to start

And that way, together, you'll be free

To take goodies from the Yum Yum tree.

Rhyme taught our first grade class by Sister Francis Marie of the Sisters of Charity, St. Augustine's Parochial School, the Bronx, New York, 1955

Introduction - Startling Statements

"Once you publish a book, it is out of your control. You cannot dictate how people read it."

Margaret Atwood

When I began speaking publicly over 50 years ago, I learned three important lessons.

First, start by telling the audience what you want to talk about so that nobody's disappointed and those who didn't come for that talk can walk out before you get started.

Second, briefly define the major terms you're using so the audience doesn't spend the time trying to understand their meaning while ignoring the point you want to make.

Third, make two or three short "startling statements" and then spend the rest of the time explaining them. Hopefully the audience, anxious to learn what you meant in the first place, won't fall asleep.

Here we go.

This is a book about information technology and its relationship to the struggle for human survival. I want to start by defining three terms.

Technology is the electronically driven machinery that replaces human labor in whatever form that labor takes.

Information technology is technology that stores, allows interaction with, and transmits data...usually using computers.

The Internet is a network that connects lots of computers world-wide. It has some sub-sets and you're most familiar with two of them: email and the World Wide Web. Some people confuse the Web with the Internet but it's actually a function of the Internet and, by the way, Social Media is a function of the World Wide Web.

Now...the statements:

1 - Among the most important questions facing the human race is "Who controls technology?"

2 - A major issue for the contemporary revolutionary movement -- the movement that is organizing the human race to save itself and move forward -- is "How do we acquire that control?"

3 - All social struggles are related to those two questions and human survival is, in part, based on how we answer them.

Ready?

The Fear You're Feeling

"That's all anybody can do right now. Live. Hold out. Survive. I don't know whether good times are coming back again. But I know that won't matter if we don't survive these times."

Octavia Butler

We can organize our survival around a slogan: "The present is ugly but the future can be beautiful."

Today, we are unified by our potential to change this world and the gut-wrenching fear we experience living in it. It's not easy to see the potential because the fear is blinding and its true cause is difficult to identify because it hides its true reason by latching onto every setback or potential problem we perceive. There are many of those.

We fear losing everything. We fear disaster visiting our family, community or country. We fear the ominous presence of catastrophic disease wrecking our lives and that of those around us. We fear walking the streets or failing at new experiences or not living up to the potential we or others believe we have. We fear not being loved or liked or respected.

We fear for the future of our children or the children of our friends.

We fear what someone might do to us and so we spend time agonizing over what the last one did. We fear white people and how deeply they hate us. We fear people who aren't white and how rapidly they are taking over our world.

We fear that, contrary to what we've been taught, there is no benign presence watching over our lives.

We fear that, when our lives are over, they will not have mattered.

We fear that, in the end, we don't matter.

In that way, our fear is a diagram of our existence drawn in the bright red ink of panic.

In confronting this fear, however it presents itself, we feel alone. That is the greatest fear we can have. Being alone painfully contradicts the most fundamental of human instincts: the need to be connected with others.

While I don't know what's scaring you today, I think it is probably a tangible substitute for the real force generating that fear: the softly rumbling sense of insecurity that we all learn to live with and that bursts out periodically, covering us in crisis. We all live it because crisis is the most persistent presence in our lives.

There is in all human culture a life that has been promised to us, presented in every waking hour of our lives as the "norm", and that promise is now rapidly falling apart.

At this point in our existence as humans, we all sense what's wrong even if don't want to confront it. In its present form, life can't be fixed. It can never return to whatever normal we believed in at some secure point in our life...if there ever was such a point. Our existence can never be what the powers of our world claim they can restore it to.

Our world is self-destructing and, if we don't change how we live and how our society functions, we will be destroyed with it.

While there are many explanations for how we got into this mess, it's pretty clear that technology has played a central role in perpetuating it and now plays a role in deepening it. Technology is not the entire force that put us on this path but it's the vehicle we're riding as we move to its end.

Yet, in the kind of irony ever-present in history and human development, technology is also the one tool we can use to save ourselves. In fact, using this tool we can not only save our world and our lives but change both in fundamental ways: making the world a place that seems impossible and reinventing life in ways that go beyond our fantasies.

4

We can build a world where perpetual fear is irrational, strife is illogical and crisis is inconceivable. Whether we do that, and how we can, is the central question facing the human race today.

The obstacles to realizing that world are immense, sown into the very fabric of our culture and consciousness.

Ours is the world of late or post-capitalism where the powerful seek to divide us by borders, religions, belief systems, gender, social role, age and movement. It's a world of imagined walls carefully constructed through manipulation of information and hiding of truth, and fiercely defended by control of communications and the repression of those who defy that control.

Revolution in this age of communications technology is not only about seizing power and production, although such a step is necessary in the process, but about reclaiming the technology and using it to re-imagine the world and our place in it.

With the juxtaposition of those stark realities -- the ugly present and beautiful future -- the most pressing question is who will control that technology: the few people who are sending us on this death trip or the great majority who want to live better?

Or, put another way and more specifically, will communications technology (the most recently developed mass technology) increase the aloneness that is the nutrient of our fear and the protector of our oppression or will it broaden and deepen our greatest weapon against that fear: human collaboration, the motor of revolution?

How do we find the hands to hold and how do we design the grasp that will endure so that, as a human race, we can confront and defeat the monster guarding the Yum Yum Tree?

The Fetishism of Information Technology

"We should not allow it to be believed that all scientific progress can be reduced to mechanisms, machines, gearings, even though such machinery also has its beauty."

Marie Curie

You can learn a valuable lesson about computers by looking at yours when it's turned off. What do you see?

It's a box containing a bunch of electronic circuits and some wires connecting them and, in many cases, there are some wires connecting the box to something else, including the Internet.

That Internet operates over a network of wires and other equipment connecting your computer with a bunch of other people's.

None of it, neither the computer nor the Internet, is doing anything until you start interacting with it.

Far too many people view information technology as a bunch of boxes and wires called computers and the Internet as this bunch of computers interacting with each other using electrical connections. We too often give our technology an independent life and view its development and history as the improvements and enhancements done to these machines.

So companies that produce new and more powerful machines have "advanced the technology" and, by extension, improved and enriched our lives. Corporate executives have magically "changed our lives" because they've run companies that produce this machinery. Our use of these computers and the opportunities they offer us is viewed as an extension of the machines themselves.

That limited perspective spawns a jaded vision of how information technology was invented and developed, how it operates and who can own and control it. Experts speak of "trends" based entirely on what these machines can do. Some warn us that the machines are "taking over our lives", limiting our social skills, chopping away at our intellectual abilities, marring

6

our ability to communicate. Others celebrate the way these machines bring us together or make us smarter or more efficient.

Well, take another look at your box. Is all that remarkable activity what you're seeing? Not at all. It's sitting in front of you, inactive and useless until you do something with it.

To attribute any real power at all to your computer, the computers of the world and the wires connecting them, is classical fetishism: "endowing real or imagined objects or entities with self-contained, mysterious, and even magical powers to move and shape the world in distinctive ways." as David Harvey puts it.[1]

The Internet is, in fact, not the technology that runs it but the people who use it. The computer isn't information technology, it's the tool you use to make information technology possible. It's when you begin your interaction with it that the computer responds and can connect itself with everyone else's computer and that's the key.

Your computer and the network it links to are nothing without you and the collaboration with other people connected to it. You and the other people who are interacting all over the world are the Internet.

That's the point to always keep in mind as we talk about what your computer does and how. Because, to do that, we have to look at the boxes and wires.

Collaboration in a Box

"Computers are incredibly fast, accurate and stupid. Human beings are incredibly slow, inaccurate and brilliant. Together they are powerful beyond imagination."

Albert Einstein

The struggle for human survival is the struggle for collaboration. Human oppression is the repression of that collaboration. Liberation is the freedom to let it occur.

If we humans are allowed to collaborate in the broadest way we can, we'll figure out a way to improve our lives. That's what makes the Internet and its technological enabler "information technology" so critically important.

Today, there are over three billion human beings using a computerized device to connect with each other on the Internet. That's developed in about 20 years. It's the fastest growing, most explosively popular communications technology in human history.

We use our computers without much thought about what happens, like we do when opening a door or climbing stairs. We don't think much about the mechanics of that door or the muscles of our body exerted in opening it nor do we think of the remarkable convergence of power and coordination involved in the stair-climbing. We just do it.

If we stop to think about these everyday actions, we realize that those muscles and mechanics are there but the act of opening the door or climbing the stairs has become so integrated into our daily lives that we seldom give them much thought. That is a significant part of their effectiveness and the reason those actions are so ubiquitous. If we had to think about how a door works every time we opened one or how our muscles work and coordinate with each other when we climb the stairs, we wouldn't open many doors or leave many floors.

So it is with information technology. Few of us stop to think about how amazing this technology is and fewer know how it works.

Yet, knowing how our computer and its connections work, at least in general terms, is important. It gives us a better understanding of its power and how it is controlled and how we can take back that control to harness its power. Most of all, we can better understand the Internet's social importance as the most advanced expression of human collaboration: which is the reason why we need to take back that control.

So let's take a crack at that basic understanding first. Don't be scared away if this seems like a "tech book" exercise. It's the last technological explanation you'll be reading in this book. The book's really about society. culture and revolution and how that relates to the Internet. But we can't really explore all that without a bit of tech background. So...

First, the computer (including cellphones which are little computers).

Computers speak and think in a language comprised of electronic code made up of strings of eight digits, all zeros and ones. That means the string of numbers is "binary" (two different numbers) and so each of the digits is called a "bit" (Binary digIT). The eight bits in a string make up a "byte".

Here's a byte: 00101101. Eight bits. Count them if you don't believe me.

The byte is the basic unit of the only language the computer understands but it doesn't mean much until it's added to lots of other bytes to make up a series. The computer "sees" those series of bytes arranged in a countless number of variations and these series of bytes tell your computer what it should do, what data it is seeing, and how it should handle that data. You may type and see words on your screen but that's the programs on your computer translating what they are seeing into those thousands of strings of zeros and ones.

For example, here's how your computer sees the phrase "hello world":

01001000 01100101 01101100 01101100 01101111 00100000 01010111 01101111 01110010 01101100 01100100

Clearly, it's a lot easier for you to type "hello world" than to memorize and type all those digits accurately. So you need a lot of translation to talk to your computer and computers are essentially translation machines. Using layers of special software programs called "language interpreters", a computer is able to accept information in various languages (including the human language of your keyboard) and transform that information into bytes its "brain" (the central processing unit or CPU) understands.

The CPU then commands various other parts of the computer to do things like turn on, pay attention to a keyboard, retrieve data stored in it, and protect itself from overload or other hazards. The basic instructions (like "turn yourself on") are on little circuits installed in your machine and the bytes you transmit to them when you turn the computer on make them take action.

Of course, you don't see any of that because the CPU is able to perform these tasks without telling you it's doing that. It can't tell you because it does it too fast to report. Like the door or the human step, explanation of how this is done would slow things to the point of uselessness.

But you do need to think about where to walk. Actions above the normal, basic and routine require processing of information and choices. The same is true of the computer. The circuits do the basics but to do more than that, your computer must interact with programs called "software".

Software: The World Visits You

"When you need to innovate, you need collaboration. "

Marissa Mayer

Computer users develop an intimate and often complacent relationship with software: the programs that actually do the things you want your computer to do. We open up a program and start working with it, confident of its responses and seldom questioning what is actually happening when it responds. In many cases, people are befuddled when the program doesn't respond as expected because, since they have no idea how it really works, they don't understand why it occasionally doesn't. For the most part, however, it does what we want.

An outcome of this relationship of predictability with our software is that we seldom, if ever, think about how it developed: who put it together, what work went into that and what is our relationship with its creators and developers? Even when some of us occasionally do think about it, many of us imagine that one or two developers sat in some room at a company and worked on making it before turning it over to the market where we acquire it.

Nothing could be further from the truth. Even the simplest piece of software on your computer is the result of a marvelous and powerful collaboration among many developers and testers, sometimes in the hundreds or, in the case of very popular software, the thousands. From the initial idea that might come up in a conversation anywhere among developers through the stages of writing code, doing initial tests, doing limited releases to releasing the full product, a computer program is the outcome of almost constant collaborative work.

In writing programs, developers will write some code, share it with others, adopt or expand it based on what others suggest, incorporate additional code written by other developers and test, retest and test again...seeking to identify problems, limitations or "bugs" (malfunctions) in the code. They put it through all kinds of uses trying hard to "break it" so they can fix anything that could pop up when the program is actually being used.

10

What's more, that work isn't limited to the people who actually do it. The involvement of software users is among the most critical component of software development. In their use of the program, a kind of mass "test" of what is often called the "alpha" version, they constantly identify problems and limitations which the developers respond with fixes.

After all these bugs are fixed, the developers produce the "beta" version which can then be released to the public with the caveat that...well...it's beta, meaning you should be able to use it for your work but let the developers know when something breaks. You can be sure that something will.

That public includes people like me and my role, along with thousands of other people, is to try it out, report problems with it, ask questions about it and make suggestions on improvement. The developers will pick up those beta comments and will often re-write part of the program to address any problems.

The beta period can last from several days to several years. As they make improvements, the developers will release new versions of that beta version and announce them so people like me can download them.

When that's over, they will make the public release which is supposed to work flawlessly but, because there is no such thing as flawless in programming, will show bugs and limitations in this "production version". So the developers make updates and upgrades often identified by number -- for example, Textwriter 2.3. That's the program I used to write this book.

That process goes on as long as people are using the program. Software that is under constant improvement and enhancement is classified as "maintained" and there are a group of identified developers called "maintainers" who read message boards dedicated to discussion of the programs, offer help with using them (when needed) and respond with their work when called for.

While most of the software you may know is sold by companies, the process just described has happened before the company released it or bought it from the developers.

Some of the highest-profile software is commercial but most software is free and open source. That means you don't have to pay for it and the code in it is publicly available so you, or anyone else, can actually change it.

What often strikes people as truly remarkable is that most of the people who contribute to the development of these programs aren't paid for their work. Even when the primary developers make money from them, many of the testers and all of the users contribute improvements and tests and feedback without making a penny. They work for the satisfaction of creating the software or because, in order to use it, they need improvements in it.

In fact, in the case of software like the Linux Operating System which runs most of the world's Internet servers (we'll get to the servers in a moment), the maintainer group is very selective. The group has to decide who will join it and when a developer becomes a maintainer, it's among the highest compliments a programmer can get: a badge of distinction.

Defenders of capitalism usually claim that their system makes possible tremendous innovation and advancement because it has managed to make human progress by ingeniously harnessing the self-interest that is at the core of the species' nature.

As contradictory as that argument is to the true history of the human race, it has continued to resonate...until you talk about software development. Something that is a literal pillar of your life is created virtually with little, if any, drive toward self-interest or wealth, based on collaboration that completely rejects individualism at every stage.

You want to build a world that is democratic, collaborative and based on collective interest? The development of software is your model and you're part of that model.

When you start using that program, you are collaborating with thousands of other people you don't know or even know about. Their thinking, work and sense of responsibility are all now a part of your life. What's more, those people have each been collaborating with many others to acquire the skills and knowledge they bring to this collaboration with you.

Software collaboration is endless ripples in the water of an intellectual pond and, after a long remarkable swim through that pond, the software is on your computer for you to use in additional acts of collaboration.

12

Software: the World Works with You

"The function of good software is to make the complex appear to be simple."

Grady Booch

The functions of your computer are overseen and directed by your Operating System, the program that oversees all the communications between all the other programs (word processing, graphic, database, etc.) and the computer's processing unit and other "hardware" (or electronic circuit) functions. The computer is like a workplace; the OS is the manager.

You know the type of Operating System your computer is using by a "brand" name: Windows, for instance, or Macintosh OS or one of the versions of Linux. People sometimes use those brand-names to describe the computer they purchase but that's misleading. The computer is just a box and wires; the Operating System is what is key.

From the very moment you turn it on, the computer's Operating System is engaging in constant collaboration with the machine's components and the programs and software you've installed on it...and, of course, with you.

Let's explain this more concretely with a quick snapshot of this collaboration at work on your own computer.

If you're reading this on a computerized device, you already began a rich collaboration: you turned it on and the CPU told all the computer's components to get ready to work as electricity flows through them. It also then automatically searched for and started talking to the Operating System asking the OS what it wants the computer to do.

Among the many requests the OS makes of the computer are: display a screen (the one you're looking at) and on it display a bunch of icons (symbols) for the various programs you may choose to use in your session. You know those icons; they're what you click on to open programs.

To show that display, the computer tells the Operating System to examine your hard drive which holds the information you've stored and the programs that work on that information. That's how it knows what programs you have available and can show their icons which are stored in the programs themselves.

The OS is also doing a bunch of other things including checking on the health of the data you've stored (what's corrupted, for example) and how much data is stored.

That amazing amount of interaction is done in a few seconds and, once it's done, the computer waits for you to tell it what to do because you haven't even started typing yet.

The moment you type on that keyboard (or use the mouse), the computer reacts. If you double-click an icon, the Operating System tells the computer to open that specific program and get it ready for you to use. When you start using it, the OS listens to the program and tells the computer what you have told the program to do: display a letter or save something or draw a line.

Keep in mind that the program you are using sends signals in the programming language it's written in which is usually a long list of human language commands and lines of code. Another program called a "compiler" receives the information sent by your program and breaks it down into the machine language the computer understands.

These back and forth communications are almost instantaneous. I type a letter as I write this and it immediately appears on my screen in a display offered by my word processing program but to make that happen the word processing program, the OS and the computer's components must perform an intricate feat of collaboration and communication, all reacting to my thinking and choices.

When I save what I've written, the OS tells the computer to store it on my hard drive. The drive's job is to store that data to preserve it because, if it doesn't, what I've written on the screen will be lost the moment I turn off my computer. The inter-action between my machine and my OS and my program is ephemeral; it only lasts as long as the computer is turned on and the program I'm using is operating. So I save it on my hard drive.

The computer is like a brain; the hard drive is like its memory. There's one element missing: the "mind". Where on the computer is that?

That brings us to the Internet.

The Internet: You're Part of the World

"The Internet, as a global system, is a network-of-networks held together by a spirit of collaboration."

Olaf Kolkman [2]

The corporations that make money from the Internet are fond of mystifying its functions but the nature of the Internet is really pretty simple. It's a network that links your computer with every other computer in the world that is using the Internet and, in the process, links all those communications going on in your machine with those going on in all those other machines.

That network is too vast for us to envision: you can't picture three billion computers (including cellphones) in your mind let alone conceive of all they are doing when they're on-line. Yet here too the nature of your relationship to the Internet is very simple. There are two phases to Internet experience: getting you on the Net and letting you share what you want once you're on it.

The first one is done the moment you go on-line. That process has become much quicker and simpler than it once was. You used to use your modem to "log on" to the Internet but, with advances in the technology, you're on it the moment you start up the machine as long as your computer is hooked into some phone line or cable. Your personal computer is connected to a box (via cable or WiFi) connected to a port in your wall connected to that big line. Your cell phone is connected through the telecom that you use for your calls and whatever else you do on that device. You pay the company that owns that line, your "access provider", dearly for that right. If you're paid up, your computer is connected to the Internet the moment you turn it on or, if it's always on (as mine is), you're always connected.

Still there? Now things get a bit more complicated.

The moment you're on-line, your access provider assigns you an IP (or "Internet Protocol") address which is a string of numbers divided by periods. As I'm writing this, I'm on-line and my computer's current IP is 142.255.81.111 -- I know because there's a service that will tell you.

That number, which can change each time you go on-line, is your identification as an Internet user. It's an id badge and all the computers in the world collaborating and communicating with your computer will know it by that string of numbers. You can pretty much ignore the number because you're not a computer but it's important to understand that every computer on the Internet uses its badge number to constantly and repeatedly identify itself to other computers.

Of course, being on the Internet isn't very exciting in and of itself. We access the Internet to do things: send email, surf the web, post our own material to the web, do research, look at pictures, communicate with others through some social media protocol. That is the second phase.

All the information you put on the Internet and draw from it is actually drawn, not directly from the other computers that are hooked up to it, but from specialized computers called servers. A server is a powerful computer maintained by the "service provider" (the company you use for email and web-hosting) and the server's role is to hold information from you and transmit it to other people who want it and who you want to have it. It's the central component of Internet collaboration.

The best way to explain how servers operate is to take a web browsing session that you're doing and probably do many times a week or maybe a day. Let's say you want to reach the website of my organization, May First Movement Technology. You open up the web browser, the software program used to "surf" the World Wide Web, and type in the site's address:

https://mayfirst.coop

We all constantly use a "URL" (uniform resource locator) during an Internet session but many of us don't really understand what it is. It's packed with critical information. The https means "Hyper-Text Transfer Protocol that is Secure" and it tells the server that you want to use a protocol (the function the computer will use) to retrieve a document written in "Hyper-Text" the language that makes links and the transfer of data. You tell it want to do that transfer in a "secure layer" which gives you some added security and privacy protection. Then you give the computer a website address with at least the domain of the site you're looking for: in this case mayfirst.coop.

In a matter of a couple of seconds, the Internet performs a series of functions that you don't see and that happen faster than the human brain can function.

Information about the URL you are seeking, where it can be found and whether your request is a legitimate one (rather than some attempted attack), is transferred among several servers situated in several parts of the world: from the large servers that store information about where the servers that hold that domain's information are located, to the servers handling the domain to the servers that actually have the page you're looking for. Each server provides a specific type of information to others and with each information transfer the search for your request becomes more specific until it reaches the server that holds the web page you're requesting and delivers it to you.

Every single web page request is a collaboration comprised of many interactions among computers, yours and various servers, that takes just a few seconds.

Try it now and see how long it takes you to get to https://mayfirst.coop.

Amazing, huh?

Congratulations if you got this far without skipping...or even if you skipped some. If all this seems overwhelming, and perhaps over-explained, you can forget the details and let one point linger because that's the point I want to illustrate.

Information technology is a dance of constant collaboration. Every single step involves some kind of collaboration among components on your computer (and you) or between your computer and others or between other computers that have nothing to do with you. That technological collaboration has only one purpose: to allow you to intellectually collaborate with the entirety of the human race.

All of what I've just explained is centered on something that is not electronic or on circuits or in programs. What makes it all work and what drives the work it does is you and all the other human beings using the Internet.

The world can now know your life, or as much of it as you want to share, and as this technology grows and becomes more popular, it becomes increasingly common for you to share more and more of your life. Others do the same. The formation of your intellect is now, more than ever, nurtured by the experiences of others. We are living collaborative lives to an extent never before experienced and that's happened because of the Internet and computer technology.

Truth be told, when I first became involved in Internet organizing it wasn't the data or information or communications with others that attracted me. As a veteran activist and organizer, I had spent decades doing all that already without touching a computer. What attracted me was this remarkable technological collaboration and the unlimited power offered by its precision and organization. As I saw the numbers of Net users grow exponentially, I noticed how the technological collaboration inherent in information technology could grow as well, modified and improved by some of the very people who were using it.

One day, shortly after starting Internet work, I had a light-bulb moment. I realized that this technology was, at once, the facilitator and the realization of the human need to collaborate and that this collaboration was extending beyond any barriers, boundaries or borders...wider than ever before. I began asking why. Why has this thing grown as it has in a world that has spurred its development while threatening our existence? I came to realize that those things, the development and the threat, are connected and they frame the challenge we face and our response to it..

A World in Crisis

"I would dare say that today this species is facing a very real and true danger of extinction, and no one can be sure, listen to this well, no one can be sure that it will survive this danger."

Fidel Castro (University of Havana, November 2005)

Today, as I write this, the most accurate description of the world is "catastrophe".

The world, as we know it and as it's currently structured, is in irreversible, irreparable crisis. Every corner of the world is ripped by wrenching human suffering.

The delicate balance between the way we live and the Earth on which we live has been upset to the point of crisis and our world climate is now the major threat we face as our extinction as a species looms as a very real possibility and an eventuality if we continue living as we do.

There is no "stability" anywhere. Every economy on earth is stumbling and the world faces epidemic unemployment and progressive alienation from productive work. People engage in increasingly desperate survival actions while the right-wing becomes more violent. Racism and sexism are becoming more pronounced and shrill. The state increases its mechanism for control and repression and has become more blatant about that. These are growing trends world-wide.

Capitalism has completely failed, morphing into an unrecognizable system in which the generation of wealth seems unhinged from the work that produces profit. If the wealth being accumulated is built by speculation and cyclical lending, the system will collapse.

- The U.S. -- with a military budget that is almost as large as all other countries' military budgets combined (almost 70 percent of the entire U.S. budget by some estimates) -- is involved in multiple wars and scores of military occupations. It maintains over 800 military bases world-wide. While sapping the country of money that could be used to address some of society's murderous problems, it also perverts the culture: lionizing military efforts and hero-worshiping combatants. In fact, the number of generals in the upper echelon of government (particularly the White House and federal agencies) is unprecedented and truly frightening.

- About a fifth of the human race has no water and all predictions indicate that that percentage will grow very quickly.

- A third of humanity suffers hunger and many scientists predict that a major food shortage crisis is imminent in the United States.

- Scientists are now saying that the human race will experience widespread disappearance within a generation and we are already seeing the first stages of that gradual extermination.

- There is no democracy in most of the world and the human race is not deciding its future.

While all these manifestations of crisis may appear isolated, they are really faces of the same crisis: the crisis of post-capitalism.

Technology and the End of Capitalism

"Capitalist production, therefore, develops technology, and the combining together of various processes into a social whole, only by sapping the original sources of all wealth -- the soil and the labourer."

Karl Marx and Frederich Engels, Capital, Volume 1, Chapter 15

The history of capitalism over the last century is the history of technology taking the place of people's work. That, under capitalism and in the hands of its rulers, has been technology's purpose. Tools, things we can hold in our hands, enhance our ability to work and produce; technology takes our place in that work and production.

For all of our lives, most of us have been guided by a description of capitalism that is actually fairly accurate. A person works, makes things, gets paid and spends that money for things the person needs, usually made by other people. In the process, this maker's employer gets to keep some of the money made selling the things the maker produced. That, we are told by capitalism's apologists, is why people keep owning things and employing people and that ownership and employment is necessary to keep things going, to keep us working and to make our lives sustainable.

This cycle of production and consumption is indeed the way classic capitalism works. In the most personal sense, it's about you working for a living and being able to buy the things you need made by everyone else working for a living. And while you do that, you make a small group of other people wealthy: the more you work and the more you purchase, the richer they get. They profit from your work by charging you (and the rest of the human race) more for what you're buying than they pay you for making it. That's how it's been all your life.

Yet it's not the way it has been for all human history. For the first time in our history, society has generalized the separation between you, as a producer of things, from the rest of your life, your mind, your knowledge, your abilities. Your creativity, interest in many topics and ideas, talents, things you've learned from others -- all of that has become unimportant in what you do to survive.

What's important to this system is the work you put into the product or service your employer is producing and that work is completely under the control of the employer because that's who decides what product and service is produced. Your specific work, defined by a boss, is all that matters. Your are your labor and, as far as this system is concerned, that's all you are.

The destructive impact of that punishing reality on our lives became clearest in the 20th Century.

Modern capitalism -- the system that Marx and Engels lived in, studied and commented on -- had developed a system of production that forced workers into a collaborative relationship exemplified by the assembly line: a product of applied electricity. During the previous thousand years, production was driven by individuals' work. Now, with the application of electricity, big machines could perform the labor of large numbers of workers and the workers' production was increasingly defined by those machines. Any individual creativity residing in their labor was replaced by a regimented enslavement to the needs and systems encased in those machines; the machine took the worker's place as the central mechanism of production.

This mix of electricity and manual labor, called industrial machinery or assembly line technology, spewed enormous value. Factories could spit out huge amounts of product using the same or fewer numbers of workers, increasing the potential profit. Capitalists could make more while paying less.

This was the industrial age, the age of electricity/labor technology. In the United States, it was capitalism's greatest hour. Here science, technology and increasingly specialized labor combined to literally flood the society with new products in ever greater numbers assisted by two huge, world wars which dramatically increased the demand for manufactured goods and paid for those goods with tax dollars creating a kind of subsidized capitalism.

As it is always is with capitalism, however, there was an undercurrent of self-destructive irrationality in all this "progress". As it produced greater numbers of products for less money, it was also displacing workers and, with increased competition for fewer jobs, it could restrain salaries. Unemployment became a fact of life in the United States and, while its levels varied depending on what our country was doing in other markets and other countries, it was always there and, for most of the period, it was actually growing.

21

This meant that the commodities produced by this new technology couldn't be turned into wealth as easily because fewer people were around to buy them. For the entire first half of the 20th Century, capitalism in the U.S. struggled with this issue: losing its footing in two depressions (one of them the most serious in its history), masking the problem through war and the production that war makes necessary but that would never be necessary in a rational society, and fighting tooth and nail battles with labor unions to force and eventually win their cooperation in this awful, abusive and self-destructive system.

Most of all, U.S. capitalism was able to export its labor exploitation and its products, easing the crisis that threatened it while exporting an increasingly brutal crisis to the economies of the less developed or "developing" countries: the majority of the human race.

Capitalism's chronic inability to meet the fundamental needs of its population accompanied by its unquenchable thirst for profit at any cost brought to much of the developing world an increased exploitation of its natural resources and a dramatic increase in factory production in some of those countries at very low wages. While things seemed to get better for American workers in the period after World War II, things actually got a lot worse for the workers in the rest of the world.

In its frenzy for profit, capitalism in this country was now producing more than its society could consume. This is the problem of over-production, the main issue for capitalism during the 20th Century and today. Technology made it possible.

Then came the 1970s and the problem of over-production became a nightmare.

Computers entered the workplace. Up to then, assembly line technology had used electrical power to do some of the work humans had been doing and organized humans to work in repetitive, numbing tasks dictated by this machinery. Now, technology not only replaced some manual tasks, it replaced most manual work and the intellect required to do that work. This technology could store massive amounts of information, mimicking the human intelligence and making many workers unnecessary.

With computer technology, the capitalist class was now able to produce a lot of its commodities without any workers at all and what were once factories full of people interacting with huge machines now morphed into barren expanses where very few workers managed complex machinery run by computers.

Today, in the United States and most of the more developed capitalist countries, there are very few industrial production workers.

As with most capitalist development, the whole thing makes no sense. There are fewer people working in this world and the ones who are working are making lower salaries and yet the rich are getting richer -- much richer. Where is this spectacular wealth coming from? It's not from more production because that requires consumption of that which is produced and fewer people have money to buy things and the money they have is less than in the past 30 years.

So how in the world are these people getting richer?

Crisis and Fiction

"The purpose of capitalist enterprise has always been to maximize profit, never to serve social ends."

Paul Sweezy [3]

In their masterpiece, Monopoly Capital, Paul Sweezy and Paul Barran identified three trends in modern capitalism: ways the system combats this problem of "over-production": the militarization of economies, the marketing of goods and services, and finance itself.

In the madness that is the U.S. contemporary economy, you can see this played out. This is a society whose major employer is the military, using the great bulk of a tax-based budget to make things that have no useful purpose but to destroy people and to pay people whose jobs have no useful purpose but to keep the thing going.

All of that is driven by a fiction: rather than a world family united by the ability to produce what it needs and the genius to continue developing that productive ability, we are actually citizens of separate "countries" mired in intense life and death competition and a natural hostility that often percolates into warfare. "Security" in this fictional world drawn for us by our ruling class is guaranteed by the ability to kill everyone else.

The goods and services we buy are marketed in a frenzy of false need and expectation, convincing us that we need things we don't, in a consumer culture of products that break prematurely and services that provide no real value. The truth hidden by this second fiction is that you don't need most of what you buy.

Finally, we have an economy that plays monopoly (the game) with real money: investing, lending and reinvesting, creating profit and value that has no real relationship to our labor or its production. In fact, most of the economies of larger, more powerful capitalist countries are driven by pure speculation -- investment in, not what a company is actually selling, but what it might sell in the future.

In this systemic casino, the average company's stock now trades for five or six times its real productive value, way beyond the profit it can ever be expected to make. Wealth, in this system, is accumulated by having stock that you paid many times too much for but that others would be willing to buy for even more.

This entire fiction, amasses obscene profit while saddling the average person with untenable debt.

We are now important to capitalism not as producers but as consumers and to fulfill our consumptive mission, we have to borrow. What most Americans have in common is that we are in debt. Like the stock market trading on speculation about companies' success, we speculate on our ability to make money in the future and be able to pay off the accumulating debt we incur today. In most cases, our speculation falls short.

This is a system devoid of real production, driven by a kind of economic simulacra (to use Jean Baudrillard's term) and guaranteed, under any logic, to crash. It represents the triumph of social fiction, imposed on the minds of masses of people.

Each one of these "fictions" -- war, marketing and finance -- is now driven by technology. For example, the New York Stock Exchange was before 1990 an arena of frenzied trading and screaming populated by between 4,000 and 5,000 people on its floor at any one time during the day and dominating the world of investment until 3:00 pm Eastern U.S. time, when the market shut down and its floor turned into a ghost-town. Today, there are usually about 500 people on the floor and the frenzy has morphed into a boring buzz.

Trading is done by computers world-wide through many different trading systems that work 24 hours a day because now, with internet technology, a trader in New York can trade in Japan defying distance and the clock.

Advertising and marketing is accomplished by technology to such an extent that many retail stores have closed or face perpetual challenge. A huge percent of consumer purchase is now done on-line through such services as Amazon, a source that appears to have every product anyone could imagine that can be purchased with quick key-strokes and delivered in less than three days, sometimes the next day. The "labor" is hidden in Amazon's warehouses and trade centers where under-paid workers (many of them working at a computer) handle an ever-increasing number of orders and fulfillments with absolutely no personal contact with the purchaser.

Finally, war is now conducted, more and more, by computerized drones and strategized and organized using computer programs. The precision of its weapons and its strategies is assured by computers.

These three industries -- finance, advertising and war -- were the places where the children of workers dumped out of industry during the 1960s and 1970s went to work. They employed most U.S. citizens either directly or through one of their many tangential industries. In the latest example of capitalism's irrationality, technology has now seeped into these industries to displace many of these new workers.

If there is one thing that defines contemporary capitalism, it's displacement. The workers of the post-war United States were most concerned about which job they could find that would offer the most against a checklist of convenience and prosperity: closest to home, better benefits, presence of a strong union, possibility of promotion. Today, workers are most concerned about job security: protection against the devastating threat of a business closing or a major lay-off. For most workers in this country, there is no job security and the checklist of convenience and prosperity seldom even figure in their choices.

The economic system under which we live is today divorced from the social purpose of an economy: to take care of its people. It's more irrational and self-destructive than it has ever been. It is going to collapse and when it does, it's going to collapse on our heads: killing masses of people, destroying the balance with nature that is essential to our survival, forcing hundreds of millions of people into migration, destroying societies as they collapse under these new pressures.

The Vaporization of the Proletarian Strategy

"For a century, the concept of proletariat anchored radical analysis and politics, theory and practice. It's now all but disappeared."

David Rosen [4]

The only alternative to this frenzied destruction is to stop the collapse by taking back the economy and the society: taking control of our society and lives.

That hasn't changed. What has changed is how to do it.

For most of the history of the revolutionary movement under capitalism, the issue of how to take back our lives rotated on one strategic question: the role of the industrial working class.

Classical revolutionary theory, starting with Marx and Engels themselves, held that the strategy to lay capitalism to rest and venture into the next stage of human economic development centered on the role of the people who worked directly with industrial production. Marx, and traditional Marxism, refer to these people as the "proletariat" and while many activists of the various left-wing movements that have arisen since the 1960s tend to view the formulation as an almost ludicrous spasm of orthodoxy, the strategic formulation is both simple and logical.

If you're going to change the society, you need to start where it produces and its most productive sector is industry. Not to argue that industry produces more profit or product or anything else than any other productive sector. It's just that, for most of capitalism's history, every other sector has depended on industry and there is no sector that rivals industry in involving human labor in its entire production. Workers are at those machines making stuff and, without their labor, no stuff gets made.

So, if you want to change the society, get those workers to take over the machines they are working on in an ordered, organized and politically-driven way, mobilized by a conscious desire to build another kind of society. Viewed

26

that way, revolution is actually pretty easy to at least start and, once workers seize this means of production, the revolutionary drive will flow through all the other productive sectors of the society. People will respond.

So you build a party, a political organization, that recruits the most conscious "leaders" of that industrial proletariat and, through that party, coordinate these peoples' leadership of the workers they spend their days toiling with. Soviet pioneer Vladimir Lenin labeled this the "vanguard party".

In the period after the sixties, many movements and thinkers rejected that strategy. In conferences and meeting too numerous to remember, I encountered opposition of all kinds to that formulation and much of it was from active, motivated and often brilliant revolutionaries.

The idea of a "vanguard", particularly when it was recruited to a "vanguard party", struck them as arrogant and elitist. Some would eschew the exclusion of so many important forces, including people of color, in the strategy. Many just felt that, given the lack of consciousness among industrial workers (a very real problem), you couldn't count on them to do anything progressive let alone revolutionary.

To some extent, this was misunderstanding. The idea of a vanguard party is merely an organizational approach to coordinating the political work of this vanguard whose existence was unquestionable. In any community, certain people lead naturally and they are naturally recognized as leaders by everyone else. It's the nature of the human species to seek leadership from people who are more knowledgeable or more capable of influence or more daring in their motivation or one of those or all of them.

All Lenin was arguing was that you identify those people and recruit them rather than build a party of self-proclaimed leaders. Despite the criticisms and reluctance of people, it was a highly democratic and inclusive strategic view.

But it didn't work. On the one hand, the reinforcement of capitalism's vice-like grip on the consciousness and culture of people in industrialized societies that Gramsci so brilliantly recognized and described made the revolutionary argument popularly unpalatable. The call he made for the development of a proletarian culture as the fulcrum of proletarian consciousness was never really embraced by revolutionary movements and those people who pushed it within those movements simply ran out of time or were run out of the movements.

27

Now there's another, fatal, problem with that strategy. As described above, there is no longer an industrial proletariat with revolutionary potential to speak of. It has almost entirely disappeared in an economy driven and controlled by technology.

The transformation of production from industry to technology poses the jarring question: if revolution means seizing control, what do we try to control? Taking over the means of production through the proletariat simply can't apply.

Part of the answer, simply put but strategically elusive, is that you seize what is producing and what's producing is technology.

Rise of the Internet and Recapture of the Mind

"Let us put our minds together and see what life we can make for our children."

Tatanka Iyotake, Hunkpapa Lakota Leader and Man of Knowledge

The Lakota people understand knowledge as the accumulation of experience and the Great Spirits as the keepers of that knowledge. The Great Spirits look over our world, take in all human experience and process its lessons. That is their role. They share these lessons with us through the person of knowledge (or "Medicine Man" as the white man called that person): the individual gifted with the ability to communicate with them, hear their guidance and articulate it.

These Great Spirits are not gods in the sense many contemporary religions view them. They don't control us; they guide us. They don't dictate the future; they tell us what that future will be like if we take the steps we should or what will happen if we don't. They show us which plants can heal. They point us in the direction of the places where food is more abundant. They help us manage the complex relationships we have with other humans close to us. They recount our history and its lessons.

They do all this by sharing with the person of knowledge the information they have accumulated through the experiences of humans who have come before us.

28

The person of knowledge hears these teachings and shares them with those who live currently. Tatanka Iyotake ("Sitting Bull") was doing just that, sharing teachings, when he called on us to put our minds together to see what future we can make for our children.

We have responded with the Internet.

While there is a strong urge to treat it as only the latest stage in information technology's development, Internet technology is radically different from the technology that has taken over production.

On its face, that doesn't seem like a very bold statement. Clearly there's a difference between a computer that runs the production in a factory and one that connects to this vast network of human interaction. It is unquestionably a new step in the process of technological development: one that has captured communications and robustly driven it to new stages of capability and speed.

Yet to view only in its impact and social functionality is to under-estimate its remarkably transformative power and, by extension, to miss its breath-taking transformative potential.

The key to understanding that power and potential is realizing just how massive the Internet actually is and how central to our lives it has become. No technology has come close to involving such a large percentage of the human race and none has so radically transformed human interaction.

Today, over three billion people are connected to the Internet, using it to conduct important activities in our lives from communicating with those we know to purchasing what we need to reporting and recording, on some data-driven equipment, aspects of our personal lives: thoughts we have and events that are critical to us that up to now would have remained completely private.

So imbued is this activity in our culture that many of us spend large portions of our days conducting on-line communications (usually with cell phones) without giving very much thought to the mechanism of that communication...or its implications for us and our society. In the sphere we have come to call "on-line", we now share our lives with hundreds of millions of other people.

Even those who don't directly use the Internet, still the majority of the human race, are profoundly impacted by it. Our lives today are recorded by governments and corporations through a massive data capture that digitizes

29

everything from where we live and meet and shop to all our government, school and personal records. Whether we share our lives on-line or not, they are captured by others and shared by them with other forces whose goal is to control and steer us.

Yet, despite the efforts to control our communications and take them over, there is something enormously powerful and liberating about this new stage of technology. Internet technology today transcends previous technology's impact on our bodies and our work to begin a process of transformation of our collective mind.

In the previous short description about how computers and the Internet work collaboratively, I compared the computer to a brain and its hard drive to memory because it's easiest to understand as a mirror of our intellectual capabilities.

We all have a brain. It's in a person's body and it functions, as do all organs, in an orderly and controlled way. It has been the subject of countless studies and books that track and describe its operations. Different parts of it play different functions in our thinking, our responses and our physical operations. There is apparently some hard-wiring because, without that, the brain couldn't function.

That's a lot like the firmware in your computer.

There is also a set of functions, many installed by the culture and socialization of a society, that guide the brain's interactions with all the organs and with our cognitive activity..

That's like software.

Then there is the memory which is key to our brain's full functionality. You can't really think without remembering. Memory is the storage device (like the hard drive) that holds what we have experienced and learned, providing us the material to process our understanding and analysis.

The brain, sitting in our heads, is both the warehouse of experience and the mechanism to process it to make it useful for our survival.

In that sense, the computer mirrors the human brain and, what's more, a computer can replace all the brain's obvious functions. Clearly, you need a human brain to program the machine to do that but, once programmed, it really doesn't need your brain for much at all except the data it processes.

30

Yet when people explore the possibility of the computer actually replacing our intellect, they hit a logical stop sign. There is something we do intellectually that an individual computer can't replicate.

We can imagine, create and plan in ways that elude the computer's programming. We can speculate in logical and reasonable ways. We can impose a morality or ethics on our activities and the aspirations that drive them.

The computer can probably do all that if we program it with the information we have but, without our human information, the computer can do none of it. Since our human information is constantly being updated, there isn't a whole lot of human functionality possible from that box.

Why can't a single computer, acting like a brain, replace the mind? The answer lies in an understanding of what the mind actually is and that is the clue to the real power of the Internet.

The "Mystery of the Mind"

"But perhaps the time has come when we may profitably consider the evidence as it stands, and ask the question...Can the mind be explained by what is now known about the brain?"

Wilder Penfield [5]

Humanity's inability to take back control of the world is partly rooted in the idea that we each have a mind that functions separately. It's a fairly recent concept, popularized with the rise of capitalism and the belief in "original thinking".

Most people believe that we have a mind and it's ours, individually -- hence the belief in "genius" as an attribute of certain few individuals or the concept of "inventor" tagged to one person or a small group of people. Most governments celebrate it, respecting the bizarre system of "copyright". Most societies honor it by attributing to their leaders the ability to conjure up original ideas, offering "prizes" for the most dazzling displays of intellect. Many religions insist on it and, in the case of Christianity, insist it is a "mystery" understandable only by God.

31

It is among the most embedded concepts in contemporary culture, which is amazing since it is patently and demonstrably absurd. Our physical experience tells us that all human active organs and organic systems must have a place in our bodies. So where is the individual mind?

When I was in grammar school, the nuns taught us that this thing called the "mind" is one of God's mysteries: that array of intangible ideas and connections that conflict with or seem unhinged from the reality we know but that God, who knows everything, easily understands. Maybe that mystery will be revealed to us when we go to Heaven to join our divine creator or maybe not. It was never made clear to me what happens if we go to Hell.

The kid who listened to that responded with a silent but very well-remembered "no". A 10 year old Puerto Rican street kid from the South Bronx needed a lot more proof than that and this idea of "mystery" (so central to Catholicism) never snared me. If I couldn't understand it, I questioned its existence. I questioned the nuns and then the Christian Brothers openly and was often ridiculed and sometimes punished for it.

When I once questioned this very "mystery" of the mind, the Brother teaching my sixth grade class encouraged six classmates to beat me for several minutes while the class cheered. That was my punishment for questioning God's humanly inaccessible understanding. They were told to only hit me in the arms and body but in the delighted frenzy of their childish sadism, a few blows to the face got in. I was then punished when I got home and my mother saw the bloody lip and black eye. She figured that, once again, I'd been fighting. All because I didn't buy into this idea of "the mind that only God understands".

Not even a beating, which a kid from that neighborhood was used to, could assuage my doubts. The question still gnawed at me: what is this "mind" people are talking about? The definitions available to me made no sense. It's an individual adventure that each of us, as adventurer, can't explain. It has no location in our body that a scientist can pinpoint. Its existence conflicts with all the science we are taught.

It seems like nonsense. Yet the idea of the individual mind persists.

For most of my adult life the question continued to plague me: why did this idea develop and what is its use? Is it, as the German philosopher Thomas Metzinger explains, a function of a mythology of "the self" or perhaps one of the persistent and powerful myths French philosopher Georges Sorel wrote about?

Or is it, as several Marxist writers argue, a product of the capitalist mythology in our current stage of human development? After all, the definition of this "mind" has shifted with the rise of every major social system and is uniquely defined in capitalist society.

For capitalist culture, the activity of the mind -- imagination, speculation, complex analysis, relating what we learn to our own lives' experience -- is a process that occurs within some individual intellectual domain, a kind of mystical realm within the individual.

Science breaks down. We can trace everything the brain does within a physical realm and even begin to explain how that activity is performed but once we get to these functions of the mind, we devolve into metaphysics. You have no idea where your mind is and you have even less explanation for how it does what it does.

The quandary leads some scientists to determine that it doesn't exist. Its purpose is only to illustrate and explain some facet of our culture.

Yet all these capitulating explanations fall short because, as elusive as the mind is, its functions are very real. We all imagine and speculate and try to apply what we're learning to our lives and our experiences. We routinely go beyond the functions of the brain. Something is happening and it demands an understanding.

The Collective Consciousness and the Myth of the Original Idea

"In our modern society we have this belief that mind is brain activity and this means the self, which comes from the mind, is separate and we don't really belong. But we're all part of each others' lives. The mind is not just brain activity. When we realize it's this relational process, there's this huge shift in this sense of belonging."

Daniel J. Siegel [6]

33

It was only when I began working on the Internet that I realized something: if we go beyond the boundaries of our individual lives and the brains in our individual heads, we can come close to that elusive understanding of the mind.

While writers like David Eagleman, whose career has been centered on the study of the brain, see its functionality in a deeper and more complex way, even they accept the idea of a functioning based on retention of information. Your brain retains and processes every split second of experience and information that is your existence and your memory is essential to that functioning.

So how does it jump to being a "mind"? Maybe we should ask a different question: "Where does the information we retain come from?"

There has never been a truly original idea. Try this: erase for a moment all that society has taught you about the hierarchy of intellect and the originality of genius. Just think about how you think. Where do the ideas you have actually come from and, if you can pinpoint a source, where did that source's thinking come from? If we free ourselves of the blinding definitions our culture has imposed and really explore the evolution of an idea, we end up going wider and wider in the human community and further and further back in its history.

Eagleman's most recent work argues persuasively that no mind can function in isolation from others and that the function of each brain is to react to the stimulus of contact with other humans. Right now, every single idea in your head, every reaction, every perspective is a composite of what you have learned from other people, even if you don't know them. Any attempt you might make to identify an original idea on your part will fall embarrassingly short upon real reflection. Your ideas are not yours alone; they can't be.

Could Einstein have posited relativity without the vast thinking and theories of his predecessors? Could they have made their contributions without the thinking of others? When the great physicist considered time and space, wasn't he taking for granted the experiences the human race has had with movement and time? Could there be a theory of relativity without humanity's conquest of forward movement or its ingenious use of space? In fact, could he have come up with those ideas if he hadn't immersed himself in vast libraries of learning, observation and theory comprised of the work of other physicists who had previously gone through that same process of absorption?

34

Einstein's thinking took place in a life that was a composite of all the human experience that preceded his. We can argue about how much his General Theory of Relativity was original, but measured in percentages, what is new and original is a microscopic fraction of what the theory actually says.

To arrive at his theory, Einstein not only relied on his own Special Theory of Relativity (which itself is not really original) but Newton's theories, the thinking of many other physicists and, finally and most importantly, the human experience observed and recorded by an entire race of living beings over thousands of years. To explain the theory, he took all that for granted because, as humans, that's what we do. But it's still there in his thinking, acting as a foundation for it and a launching pad for his speculative theoretical adventures.

The genius of Einstein's mind wasn't some invented "originality" but the ability to take in a vast amount of experience and order it productively in order to extend it. In other words, he "got" more of the human mind than most of his contemporaries.

That's how our brain works -- taking in experience and synthesizing it because all of our thinking is based on reality as our species knows it. All the thinking that has gone into building that human knowledge over our entire history is contained in our individual thinking. Every idea in your head is, in some way, a contribution made by the entire human race.

That's what Tatanka Iyotake, a master of precision in language, meant by "put our minds together". The individual mind may be a myth but the mind is the very real, mass collaboration of the thinking done by the brains of the entire human race and the experience that drives that thinking, collected in thousands of years of recording and reflected in thousands of cultures. It is the greatest act of human collaboration.

For the entire history of capitalism, and most of the systems that preceded it, dominant forces have denied us this revelation. They have done it through the popularization of their philosophy and their divisive, stumbling perspective on thinking. You can see them stumble when they try to characterize thinkers.

Take Karl Marx. Is he a sociologist -- well, he's considered one of the three "founders" of modern sociology. A philosopher? Well, he does pretty well taking on Hegel and Feurback. An economist? Well...he wrote Capital, a work that has influenced every economist since its publication. A writer? He wrote a lot and lots of what he wrote is published. An organizer? There's a laundry list of organizations and networks whose creation he and Engels led. We're running out of space.

Those same types of questions can be asked of virtually every influential thinker. People can't really think in disciplines although, forced by social convention, they may try to write in them.

The indigenous cultures could never fathom this division in that knowledge because human experience cannot be divided into disciplines and it's that experience that the Great Spirits absorb and share. You don't live history or sociology or science: you live all of that and more simultaneously and your daily experience is shared, through knowledge, by the experiences of everyone on earth.

Once we free ourselves of this enslavement to knowledge separation, the idea that the knowledge in our heads is separate melts away and reveals the truth of the mind.

Is there anything going on in your life that isn't affected by the explosive changes in our environment or the advances in our sciences or the new thinking about human behavior? Does that not affect your thinking about every single thing you think about? And is not the idea of human survival as a goal -- a concept whose universal acceptance is fairly recent -- driving so much science?

Where is all that coming from? From other people and their experiences, some of them unfolding in faraway places you've never seen, much of it deriving from ideas and experiences that precede ours. The ideas in our mind are inherited, absorbed and refined over and over in a process that is framed and influenced by the conditions each generation of our race survives and seeks to control. That survival and control is made possible by the changes we make in our lives and our world based on our imagining of what can be. We have survived as a species because we can see a world tomorrow that is different from the world today.

Our brains understand the present; our minds imagine the future. That imagination is a collaboration of all humanity.

It's that very function of the mind that capitalist culture seeks to quash. You can feel that repression, although it happens continuously and so you may be used to it. It's that push against trying anything new, the feeling that nothing is going to work and the constant nagging feeling that at any point everything could go bad. It's the belief that "imagination" is somehow contradictory to knowledge and so should be relegated to a small portion of your thinking...if at all.

The issue is brought home when we think of the problems that threaten us and could be solved by using what we already have and reordering and reorganizing it. Yet that is difficult to do because it would take imagination and imagining is imprisoned in this society in the myth of individualism.

In fact, those of us who imagine a different society are viewed, derogatorily, as "dreamers" or "impractical". The prejudice against us is so deep that it affects, not only what people say personally, but the ability to find work, to be heard in the public discourse, to hold political office, to be socially respected. Those who speculate about what society could look like were it to fulfill our needs are viewed as fundamentally irrelevant to any real social planning or political decision-making.

Despite that intellectual repression, this kind of thinking is going on in minds all over the world in rebellious reaction to the desperate state we are in. Were we to recognize the power of that collective imagination, things could radically change. What if political and social discussion not only included imagining but were centered on it? What if the goal of our political thinking was, not just managing what is, but drawing a picture of what would be? What if politics were about "putting our minds together" about the world we make for our children?

In the Technology and Revolution sessions May First Movement Technology sponsored in the United States and Mexico, about 1500 activists converged in nearly 25 sessions of imagination. We asked the participants to use short phrases to identify components of a society they would like to see. We wrote those ideas on a board and by the end of the 15 minute exercise, four or five boards were filled with ideas...the most extreme and liberated ideas people could come up with.

Yet when we examined the board, we realized that rather than some flight of fancy, these ideas were all feasible and grounded in real human experience and expectation. The world on those boards was possible.

Were we able to collaborate on a meshing of our minds in our design of our world, were the process of imagining viewed with respect and true consideration, were the human mind the real definer of human activity, there would be no real barrier to social reconstruction. We would all want to make a revolution because that is what our mind is telling us to do.

The Genius of the Internet

"It is the long history of humankind (and animal kind, too) that those who learned to collaborate and improvise most effectively have prevailed."

Charles Darwin

Viewed as a meshing of minds, Internet Technology isn't just a new step, it's a complete departure. That's sometimes difficult to grasp because its monumental impact isn't immediately apparent. The Internet isn't the first technology to provide communications; the telephone does that and that's only one example. Certainly it's not the first technology to offer information or "intellectual expansion"; no technology has done that better than television.

Yet Internet technology facilitates, even makes necessary, a series of interactions that are unprecedented on such a mass scale, intellectual depth and transformative potential. With its billions of users, each contributing thinking, experience and information and each drawing from the information of all those others, it captures the process of the mind and becomes a vehicle for the mind's activities.

In my contribution to the book The Organic Internet, I defined this Internet as primarily a massive movement of humans whose purpose is to communicate. At once, it was an obvious assertion and a controversial one, pushing back against the view of the Internet as a technological phenomenon. Certainly, there is technology enabling it and there are people who insist that the Internet was invented by a group of scientists, working for the military, who developed its complex protocols. The problem isn't that such a perspective is wrong but that it doesn't tell us much about the Internet or capture what the it is today.

Viewing the Internet as a mere invention of technologists is like saying the developers of gun powder invented war or the people who first developed the hammer invented housing construction or, for that matter, that those who refined the concept of housing invented cities. These greater developments are social developments, outcomes of collaboration by masses of human beings using tools and technology.

To go a step further, nobody could argue convincingly that a city is merely a group of structures built with tools. That would ignore the broader reality of this incredibly complicated and powerful human invention. The very concept of a city involves not only a social contract based on collaboration but a centralized, shared relationship to the production and circulation of goods combined with bold and even risky collaborative imagination. As our cities developed, in alternating small and explosive increments, human beings were applying solutions and aspirations borne of the imagination.

The same is true of the Internet.

If we start with our digital "brain", the computer, and its memory, the hard drive, we can trace the reality of the Internet in a way that transcends the electronics and physical structure of the system based on the transfer of those "bytes". The "bytes" carry the information we share. At any moment of the day, billions of them are rushing at unprecedented speed, carting information about your life experiences, the reactions you have to them, and your analyses of them to countless interested people while you take in all that they want to share with you. Without that information, the bytes have no purpose or meaning.

Technologies don't create the needs these developments address. It's the needs that create the technologies and the Internet was created by the need of the human race, confronting its imminent extinction, to communicate and collaborate universally: to share the mind.

The key to all analysis is the question "why?". Why do so many people use computers and cellphones to communicate? Why have we, as a human race, integrated this technology into our lives so deeply that its explosive growth surpasses that of any other previous communications technology? Why, in the last 20 years, has the Internet transformed from a sparse place inhabited by technologists and some highly sophisticated and privileged people to the single most dominant form of human mass communication we have ever known, surpassing the telephone and then actually taking it over to become the dominant phone technology?

Why did it develop so quickly and, most importantly, why now?

Only one answer really makes sense: survival and the drive to thrive.

The Survival of the Connected

"In a time of crisis, the peoples of the world must rush to get to know each other."

Jose Marti

When scientists analyze our development as a species, they often dwell on the central question in our relatively short history: how did we manage to survive and thrive? Compared to so many of the animals who share this earth with us, we are small, physically weak, slow in motion and in action, limited in our ability to adapt our bodies to physical surroundings, susceptible to heat and cold. No intelligent wager could ever be placed on our survival and no logical mind could ever predict our absolute dominance of the environment we inhabit. What do we have that other living things lack?

We have survived, in large part, because of our ability to think and collaborate. No individual among us could ever survive in this world alone. Our strength resides in part in our ability and drive to form communities which can think collaboratively.

Yet, while that explains our survival, it still doesn't explain our dominance. Many species collaborate in ways that dwarf ours. There is no greater collaboration than that among bees, for instance, in which an individual is not only a contributor to the collaborative progress but actually a slave to it. Animals working in packs collaborate as do animals, like birds, who group instinctively for some specific task like migration.

Nor is our ability to think much of a distinction. Recent science tells us that other species like whales, dolphins, cats and apes, think all the time in ways more complex than we ever imagined.

Finally, our ability to communicate isn't the key difference because all animals communicate, some in ways whose sophistication and complexity we have yet to truly understand and appreciate.

The key difference is that those living things that share our planet think and communicate in strict reaction to what surrounds and affects them right now and their collaboration flows from that reaction. They are, effectively, constrained by the reality that envelopes their lives. Their methods, tools (such as they are) and socialization are all based on an immediate perception

40

of what is and, as far as we can tell, they don't go much beyond that. They are reactive. They lack imagination.

We, among all species, imagine a future and our survival is based on planning for it and improving in it.

For virtually our entire history, we have been recording our history: first primitively in drawings on walls to more sophisticated tracts and writing and now through moving images and recorded sounds. We do that because, in our consciousness, we realize that there are important lessons to be learned from what happened today and yesterday and what we did in response to it. We know that those lessons may prove important as we envision and plan for the future and we understand that that future will be full of challenges and possibilities that erupt from meeting those challenges.

We, more than any other species, can envision a tomorrow. We survive as a species because we have imagination.

This imagination isn't speculation. It's not a flight of fancy. As the Technology and Revolution sessions demonstrated, it is grounded in our understanding of what is and has been. That's why history and imagination are so tied together. The sharing of imagination requires an absorption of the past and an appreciation of the present. The more information we can exchange with each other about what has happened and is happening, the more robust and encompassing is our imagining.

Throughout our development, all that we do socially -- from communication to recording of information to organized sharing of knowledge -- is performed in service of our imagination. From decisions about where to migrate among our first ancestors using their understanding of available food and safety of conditions to determine what living place might be better to the complex and data-driven decisions we make today in crafting and constructing our collective lives, we use what we know to project what could be.

Our development, in fact, can be mapped in the growth of communities, both in size and complexity, and those communities can be described as arenas of shared imagination. What we call culture, our use of the tools of our survival, is sustained and nourished by our collaborative imagining. Successful societies managed to collectivize imagination from larger and larger numbers of people.

41

Now, for the first time in human history, we can share that imagination with the entire world and avail ourselves of its creativity and power.

To understand this marvel we call the Internet you have to go beyond its wires and machines and protocols and bring into relief what it does, what we expect of it and why.

Faced with relentless oppression and possible extinction while we have the ability to quickly solve all our major problems, the human race has turned to the process it has always used to survive and thrive: communicate what we experience to share our imagined possibilities. We seek from the entirety of humanity a sharing of its collective imagination and then seek to collaborate to make it possible.

That seems so logical and desirable. So why have we not been able to do it?

The Empire Strikes Back

"We can sure of two things. We will struggle for our freedom and those who oppress us will do anything to make sure we don't achieve it."

Juan Mari Bras, Secretary General, Puerto Rican Socialist Party, speech in
New York City 1974.

One of the primary themes of the epic and apparently self-perpetuating Star Wars franchise is that the advancement of society through struggle provokes an inevitable reaction: the forces of oppression will always strike back. Just when the forces of good led by the remarkable and often under-rated Jedi scored what seemed like the final victory, here comes the Empire with something new, more powerful and devastating.

If we view the Internet as a mechanism for the meshing of minds, it's pretty logical that capitalism would quickly seek to take it over to quell its danger to the system and take advantage of its innovation. The system's drive for Gramscian hegemony moves it to violently consume every aspect of digital communications and re-frame it as a tool for profit and control.

The ways it has done this in a matter of a very few years are multi-faceted, matching the multi-faceted nature of our culture and communications, but there are three silos in which we can fit this repressive effort: the surveillance state, the commercialization of communications and the increasing control over your life.

The Surveillance State (Surveillance Part 1)

"The right of the people to be secure in their persons, houses, papers, and effects against unreasonable searches and seizures shall not be violated, and no warrants shall issue but upon probable cause, supported by oath or affirmation, and particularly describing the place to be searched and the persons or things to be seized."

Amendment #4, The Constitution of the United States of America

The constitutional right to privacy has long been a pillar of protest and organizing for social change. The Fourth Amendment frames a stunning power we have over our lives and society: the right to information of our own and the right to protect it and share it with anyone we want without the government seizing it or listening in on those communications. Without that right, and our movement's fierce defense of it over decades, we wouldn't have an opposition movement in this country.

Today, that right is no longer respected. At this point, everything you do, write and say is probably recorded by someone and stored someplace for the use of the government and its policing apparatus. What is known as the surveillance state is among the most successful and best-known government and corporate "push-backs" against the liberatory use of the Internet.

It's a two-prong push-back. On the one hand, changes in federal law make expanded searching and data capture legal: slipping in between the lines of the Fourth Amendment in an increasingly frequent and brazen feat of legal acrobatics. On the other hand, much of the data is captured by companies who later transfer it to governments and that corporate surveillance isn't illegal anyplace on Earth.

The roots of surveillance grow deep in the history of this country. My first contact with it was the relentless spying done the Puerto Rican Socialist Party and the periodic and unfailing visits by FBI people to my place of work and my home -- threatening me, trying to get me to answer questions about my comrades, taunting me about my silence. Statements I made at meetings would turn up in government reports. People I knew (including family members) would get calls from the FBI asking if they knew of my "terrorist activities" or what places I most frequented or who my friends were. My father suffered a heart attack after FBI agents visited his home to ask about my "terrorist connections".

The surveillance was also personal. When I made phone calls I would frequently notice a quick "tap" or a sudden change in sound environment (as if I was in an echo chamber); I learned that this was an indication of an active remote tap. And they were always there -- two white men in suits standing across the street from home or office, staring and doing nothing else.

Other revolutionaries of the 70s fared much worse. Many organizations were literally ripped apart by infiltrating FBI undercover agents and many people (many of whom I knew personally) had their lives ravaged in a trail of lost jobs, broken relationships and imprisonment for refusing to talk to Grand Juries.

Surveillance was ever-present. We were confident that we were being watched at every meeting and had to take measures to meet in secret, usually in a different house every week, to avoid the watching eyes and ears. The horror of it all is that, as documents show, we were underestimating the level of intrusion, disruption and surveillance and the enormous amount of human resource dedicated to it.

The recent history of the capitalist state -- the part of society tasked with keeping the status quo intact -- is a mix of consistency of purpose and constant change in method. The purpose is, for the most part, to repress movements that challenge the social structure and mobilize in a revolutionary way. That's never changed. The methods, however, have undergone spectacular change.

The growth of on-line communications has produced the realization of the repressive state's dream: the ability to capture virtually all activity by almost everyone in a society. The data of our lives is maintained on giant computers that draw it from all kinds of sources and, through sophisticated software, match the various pieces to draw a full and comprehensive picture of who we are, what we do and what we want.

Take an average day. You wake up, get ready to face the world and emerge from your home. Some people watch a bit of the news on television. If you do that, the signal provider has recorded what you've tuned into.

The moment you leave your house, street or gathering-place cameras will pick up your image and some will actually record a video of your movement through its viewing zone. At some point, you're going to buy something and, if you use a credit card, that transaction is recorded and stored. If you pay cash, you may have to draw the money from a bank and that's recorded and stored. Throughout the day, all your home activities can be monitored by one of the many "personal robot" products -- like Siri and Alexa -- that not only do what you want but record all that you've requested.

The moment you use your cellphone (and most reading this probably have one), your location when you used it, the numbers you called, any data connection you have made (including web-browsing or texting) is now in the data storage banks of your phone company.

You use your computer to go on-line -- every website you visit is tracked and stored and, if you use an email provider like Gmail, your email (sent and received) is stored.

All the data in your wallet or handbag -- driver's license, credit cards, medical insurance cards -- is stored as captured data, as are the details related to how you use them.

If you're in school, your records are digitized and stored. If you have been arrested or done time, the stored information on you runs several pages. Public assistance? They have records on the times you've visited the welfare office, your family, rooms in your home and the payments you've received.

Everything you do may very well be recorded.

As horrible as all that is, it isn't the main problem. All that information is, after all, in lots of different places and, as long as it stays in those places, it's intrusive and violating of your rights but it's not a full profile of your life. The problem is that modern police and government technology is rapidly developing the ability to collect all that information from all those sources and synthesize it to develop a detailed profile of who you are and what your life is like.

In massive data centers, called "fusion centers", the government collects all this information. Purportedly set up as data collection centers to support the war on terrorism, fusion centers have quickly morphed into banks of personal information about millions of U.S. residents. They draw the data from law enforcement data bases, private and corporate sources (like Google), the military, and just about any other source they can identify.

That data draw is not necessarily direct -- a fusion center might not get footage from a street camera in your city, for example, but that kind of footage might very well find its way into a police department and that will almost certainly go into fusion. You need not be doing anything other than walking down a street; the police may have captured footage of that street for several hours if a suspected crime might have taken place in a couple of minutes. But they'll turn all the footage over to the center.

The same is true of much of the information collected about you. Even the data collected to share with advertisers, like the tracking of your information requests using Google (that is almost immediately reflected in the ads you see when you go to that service's search page) can and will be turned over to any government agency seeking it through subpoenas.

In most cases, nothing much will come of this. In fact, the Center staff probably won't even compile your data if you don't pose a threat to the government. But the problem is that it's the government that decides that you're a threat. So any communication you might have on some political issue, any demonstration you might attend, or even a personal contact with someone who is a surveillance target (even if you don't know the person's activities) will trigger a fusion of your data. At that point, everything they have about you is unified and organized.

The data capture is designed to identify, investigate and combat people who are "threats" but who defines who's a threat? David Rittgers of the Cato Institute, a conservative think-bank, sounded this alarm in 2011:

"This follows a long line of fusion center and DHS reports labeling broad swaths of the public as a threat to national security. The North Texas Fusion System labeled Muslim lobbyists as a potential threat; a DHS analyst in Wisconsin thought both pro- and anti-abortion activists were worrisome; a Pennsylvania homeland security contractor watched environmental activists, Tea Party groups, and a Second Amendment rally; the Maryland State Police put anti-death penalty and anti-war activists in a federal terrorism database; a fusion center in Missouri thought that all third-party voters and Ron Paul supporters were a threat; and the Department of Homeland Security described half of the American political spectrum as 'right wing extremists'." [7]

If you're reading this, you're a target.

In fact, you can become a target of this big brother surveillance by doing nothing more than being of a certain demographic or in a certain community. Today, police departments in about a dozen states in the United States have begun using "predictive policing" software to organize and plan police deployment and to ready their police for response. The idea is that this software can handle and analyze data on criminal behavior better than any "human driven" method and so it's more useful to the law enforcement department.

The problem is that the data it uses is often packed with the biases and prejudices of the cops doing the arrests and stops that make up the data. "When you feed a predictive tool contaminated data," ACLU's Ezekiel Edwards writes, "it will produce polluted predictions."

What's more, even data that may be accurate doesn't make police more effective in protecting communities. "If a police department places a premium on over-enforcement of low-level offenses over reducing communities' entanglement in the criminal justice system," Edwards writes, "or if its mindset is characterized by militarized aggression and not smart de-escalation, or if it dispenses with constitutional protections against unreasonable searches and seizures and racial profiling when inconvenient, then predictive tools will only increase community harm." [8]

The obvious threat here is to your privacy and your ability to exercise your rights: essential to building a better world and surviving in this one. But there's another threat, almost as pernicious although less obvious.

State Surveillance (Surveillance Part 2)

"The man is wise...and you're still an outlaw in his eyes."

Steely Dan -- "Kid Charlemagne"

The public conversation about this intrusive use of technology has been mainly framed as a conversation about "citizen rights". Organizations like the ACLU, one of the leaders in the struggle against this surveillance abuse, argue for a "balance" between abuse of rights and the right of the government to "police" us. Here's the ACLU on those fusion centers:

"There's nothing wrong with the government seeking to do a better job of properly sharing legitimately acquired information about law enforcement investigations - indeed, that is one of the things that 9/11 tragically showed is very much needed.

"But in a democracy, the collection and sharing of intelligence information - especially information about American citizens and other residents - need to be carried out with the utmost care. That is because more and more, the amount of information available on each one of us is enough to assemble a very detailed portrait of our lives. And because security agencies are moving toward using such portraits to profile how 'suspicious' we look." [9]

Well, yeah, but the government of the United States isn't a neutral arbiter or a referee in some contest over social power. It isn't there to defend our rights or protect us from others or help us attain a better life for our kids. It's not our friend. It's an institution whose primary purpose is to defend the interests of the powerful group of people who control this society. Part of that job is to repress and, if necessary, destroy anybody who challenges that group's power.

In the debate over surveillance, that point is too often lost. We should allow no surveillance because the people doing the surveillance will use the information to repress us and even kill us. Giving somebody a right to examine your life so they can kill you is pure insanity. That's the point made by "surveillance abolitionists" like the Stop LAPD Spying Coalition. Whether the government's collection of information about you is necessary for your protection is, in fact, an irrelevant issue because the government isn't doing it to protect you.

It becomes harder to convince people of that abolitionist perspective, however, when surveillance is such an embedded force: a part of our culture and a deep-seated part of your daily life. How can you get rid of something that is part of your routine, reactions, actions and involvements every day? Not only does this surveillance support oppression and repression through collecting data that is needed to repress but it supports that oppression culturally by making you feel impotent and incapable of changing the situation.

Like many technology activists working in collaboration with non-technologists, I've frequently confronted the attitude of complete resignation, bordering on nonchalance, from activists who accept blanket surveillance as an unalterable fact of life.

Given this blanket surveillance, this attitude is understandable. But think of what it says in the long-term? If the government is allowed to surveil us all the time, we are never going to successfully defeat it. This pessimism about the prospect of surveillance abolition is effectively a surrender to defeat and its perniciousness doesn't end there.

The surrender to defeat is a nutrient of the attitude that we can't defeat this monster in the long term, a defeatist attitude that still plagues our movements: if we can't stop surveillance, how can we stop oppression?

This isn't to say we don't fight for protections from that surveillance or pressure that government to abide by them. That work can produce real results, protections that are very real and actually vital if we're going to continue our work. But, in the end, they are compromises which will prove inadequate if we don't go further.

Forcing the government to do surveillance in a "democratic way" surrenders to the government the right to surveil us and compromises the protections we've momentarily won. Like any reform, these protections can be taken away and frequently are. As long as the government has the power to surveil us, it has the power to increase that surveillance.

That's why we have to see those "reforms" as steps in a longer-term process, forcing small contradictions to government power, to reach the final fix of the real problem. The problem isn't that the government surveils us, it's that we have the type of government that has to because it protects a system that exploits and oppresses us.

The Commercialization of Communications

"The most valuable commodity I know of is information."

Gordon Gecko -- Wall Street (the movie)

Even today, with so much politically-precise Internet activism going on, leaders of our movements view the Internet as a product or service to be shopped for and purchased. At May First, which is among the best known left-wing internet organizations in the United States, the most common question we get from prospective members is: "How much does your service cost?"

That confusion in one person can usually be dispelled with an explanation and conversation. But that same mind-set about information technology is epidemic within the larger community we call our "movement". It spreads through thousands of movement conversations occurring each day, taken for granted by these activists, implicit in their frame and lens. If we aren't having that conversation as a movement, how can we have it with the rest of society? At this point, the ruling class is winning this contest.

You can test that statement about attitude by testing yourself. When somebody mentions the Internet or some other information technology tool, do you think about its origins as a product of intense collaborative creativity among technologists, its rise as the uniting of millions and then hundreds of millions of people and the way it spreads information to combat all disinformation? Or do you think of a service that delivers rapid communications and information to you and from you?

Do you think the Internet is something you and the rest of the human race has developed and has a right to control or do you think of it as a powerful and useful service provided by corporations?

The commercialization of the Internet is, at once, a coup of marketing and ideology.

Through its capacious control of information technology, contemporary capitalism has managed in a very short time to capture the purchase of goods in most larger economies -- the realization of value about which Marx and Engels wrote -- resulting in a scorched earth obliteration of most large retail businesses (mainly stores), the shrinking of face to face (staff operated) banking and finance, the transformation of a vast sweep of consulting and assistance services and the amalgamation of records and data customarily used in business and retail transaction. Your life, so riveted in the business transactions you perform, is more controlled than ever before by corporations.

Yet the most damaging aspect of this commercialization is cultural and ideological: the resulting shift in attitude people have towards their own work.

When you sit down at a computer, you are working. Technology replaces a lot of the work you would normally do to complete the task you're involved in but your labor and thinking still guide how this technology does that. You make choices, plans and decisions about what the goal is and how it's to be accomplished: whether that is writing a short email or making a purchase or crafting an extensive essay or developing a list of people or things you need for your task. You are the operator and your creativity is at its most productive.

Yet you probably don't think of yourself that way. Rather, in your mind, you are a consumer. It's not about the choices or plans you have, it's about what you need and buy or what service you pay to stay on-line or how fast your corporate-supplied connection is. In your mind, you are not in control; the control is in the hands of the corporation that puts you on-line, offers you the means to do something or sells something to you when you need it.

In his Economic and Philosophic Manuscripts of 1844, Marx mapped the "estrangement" or alienation of the worker under capitalism, the root of the excruciating pain of our contemporary lives. We're alienated from our product, our role as a producer, other workers and our "species essence": our ability to see our work as outside ourselves and visualize its form when it's finished. We can see the future. We have imagination. Yet capitalism neither values that imagination nor compensates us for it.

Now technology and capitalism have forced upon us a fifth form of alienation: estrangement from the very process of production, the idea of us as consumers.

So profound has the notion of you as a consumer been implanted, that most people can't even conceive of running an Internet without corporations. In fact, most of us can't conceive of doing any tech-enabled work without paying some company to allow us to do it.

In the movie "The Usual Suspects", actor Kevin Spacey's character (Kaiser Sose) quips: "The greatest trick the Devil ever pulled was convincing the world he didn't exist." Well, the greatest trick capitalism has pulled is convincing us that we are primarily consumers.

In that generally accepted distortion we are viewed, and view ourselves, as active only at the very last stage of the productive process that keeps us alive and that would never happen without our labor. We're merely purchasers of these objects which are presented as the product of some near mystical event. We seldom think about where this stuff we buy came from or what our role, individually and collectively, was in producing it.

The myth of the consumer severely damages our ability to think about how to fundamentally change the society. At the mercy of the mystified processes that bring these products into our lives, there's little we can do except buy them or choose not to. Our ability to take over those processes that we as a human race initiate, and which are reliant on our creativity and work, enters the realm of fantasy. We just can't conceive of assuming their control.

The Internet, this massive network we have developed through our use of its technology that would completely collapse without the participation of the three billion or so human beings using it, is the world's most intense and frequent consumption. Capitalism has convinced us that, rather than its creators and sustainers, we're only its consumers.

What's more, we're now products. The Internet's unique capture of information allows corporations to form a living portrait of each of us that's worth a lot of money to lots of companies.

People's dependence on Google, for example, allows that company to track patterns of interest (through your searches), personal friends and contacts (based on who you email with), what you buy (through your response to ads on the site). With that in hand, the company can amass a profile of you which then allows other companies to target you in their marketing and governments to trace your on-line activity through all the various surveillance programs they employ.

Social Networking, by far the most popular use of the Internet today, is little more than a data capture environment. The companies that offer the services, particularly Facebook, host your site and control all the information on it. When you think about what that information says about you, the control is disturbing. CNN writer Julianne Pepitone called Facebook, "one of the most valuable data sets in existence: The social graph. It's a map of the connections between you and everyone you interact with." [10]

Not only does Facebook hold your personal data but the personal data of all the people you designate as "friends". In many cases, displayed photos show their faces (and yours), the faces of those they come into contact with and the places where the contact took place. There are also long strings of thoughts, comments, reports on what you're planning to do or what you did and who you did it with. A single web page offers a profile of your life, your activities and your thinking. What's more, because others "comment" on your Facebook pages in an informal gathering of minds and contacts, those connections are condensed.

This amalgamation of information isn't evil in and of itself. In fact, it could be remarkably empowering. The problem is that all of it is in the hands of one large company and that company owns it, uses it in marketing studies and advertising profiles and will turn it over to any government agency that asks for it. You have no control over that. It's in the user agreement. It's published and it's no longer yours. It belongs to Facebook and anybody Facebook wants to share it with.

You are now a commodity bereft of all value emanating from your work and life. With most stuff you buy, the price is at least partially riveted to the cost of producing it which involves, in part, the cost of the labor. But with your life as a commodity, the cost (the price of you) is based entirely on your own potential consumption of commodities.

Your identity as the consumer is now in itself a commodity and, by extension, your only value as a human being.

You're nothing but a product and so not only should you not assume control of your life, according to capitalist culture, but you have no right to an independent life at all.

The Control Over Life

"Your culture will adapt to service us. Resistance is futile."

The Borg -- "Star Trek: First Contact"

Today, in the United States, your life is largely in the hands of five companies. Even if you don't know much about technology, you know their names and, more or less, what they do: Apple, Amazon, Facebook, Google and MicroSoft.

You need information, you "google" it -- so prominent is that company that its name is now a part of the English language. Depending on what you find out, you can purchase something (with Amazon) or share the information with friends on Facebook or use your iPhone from Apple to call or text or otherwise communicate with someone who probably uses a Microsoft product on their computer. You do that many times a week. We all do.

Information technology has blurred the lines between corporations and the government to the point that, for practical purposes, the line no longer exists. For instance, when the U.S. government decided to unify all its military data in one cloud storage system, it awarded a $10 billion contract to Microsoft to bring the plan to fruition. Governments paying capitalists a lot of money for things isn't new, of course; it's part of the function of government budgets. But a corporation like Microsoft in control of all military data is very new and very dangerous.

Even more dangerous is the increased centrality of data storage in corporate information technology. The recent trend in Amazon's activities is a dramatic case in point. While the company is best known as the on-line retailer of choice, its on-line sales don't make it all that much money. In 2018, the company sold a lot of product: about $208 billion in sales. However, it only made a profit of $5 billion on that retail operation. That's because it thrives on under-selling the market and squeaking by on low profits.

For most of us, calling $5 billion "low profit" is absurd but the company actually made more money in 2018 from its web services business, Amazon Web Services which offers cloud storage service to companies and governments. In 2018, AWS made $7 billion off just $26 billion in sales. As it moves toward greater control of the cloud market, of which is already owns about 30 percent, that margin is sure to grow.

The profit margins, however, don't tell the most socially significant story: most of that stored data is profile material about you. Through its storage business, Amazon is accumulating a massive profile of everyone in this country, controlling the stories of your life and everyone else's.

While hold that stored data, Amazon collects its own through your purchases, etching a profile of your life, needs and preferences.

The importance of that data is underscored by Amazon's advertising sales, motored by the personal data it collects when it sells you stuff. As Whole Food Market CEO John Mackey writes, "Amazon has been aggressively investing in digital advertising. Based on management's comments, it's already a multi-billion-dollar business -- most estimates peg it at around $10 billion in revenue in 2018." [11]

That's a business that grew by 90 percent over-all in 2018 and its impact on the culture of the society is enormous.

It insists that the cloud stored data is kept separate from this advertising profile data and, for now, that's probably true. But all that data is under its control and, given the history of corporate technology, can we trust that the separation will always be respected?

What's more, the "data share" with governments goes beyond advertising and consumer surveillance into the enforcement of repressive and draconian laws. Among government-cooperation activities, Amazon has begun collecting and turning over data among non-documented people in this country to the U.S. Department of Homeland Security's Immigration and Customs Enforcement: the anti-immigration police force. That level of corporate/government cooperation on what is among the most controversial federal government policies places Amazon on the side of repression and police state behavior.

The lesson: data, not physical products, is the current and future money-maker for this company and you are the data. The retail business is almost a loss-leader for Amazon, a kind of advertisement for its near absolute domination of the consumer culture, a source of unprecedented profile material and a lure to other companies and all kinds of governments to bank their data with Amazon.

The spectacular power of these companies, however, is not limited to the tasks or capabilities for which they've become known or from which they profit. While profits certainly drive their plans and efforts, the outcome of those plans and efforts is the control over the culture and functioning of the human race.

On the one hand, they are mapping our future. Google's development department (a sprawling system of its own development staff and scores of sub-contractors) is expanding an already huge laundry list of futurizing products and capabilities: from accessing a cell phone without holding it to producing air-borne surveillance devices to monitoring your life through "personal robot" products like Siri and Alexa.

Facebook, besides its own huge data-collection activities, has been increasingly successful in its attempts to bring a truncated and commercially-driven version of the Internet to Africa and other developing areas allowing access to only certain services and blocking access to the Net's more creative and informative functions.

In all these cases, the end-result is that the primary power of the Internet, the power to choose what you say and hear, is obliterated; a company is deciding how you plan and conduct many aspects of your daily life all the while watching and recording you doing it.

At the same time, these technological developments are aimed at zapping the last vestiges of personal social power from technology, bringing to its end a process of personal disempowerment that began almost immediately after the Net's launch. Capitalism doesn't plan futures, it tumbles into them and this future of complete disempowerment is the future this system is tumbling into, taking all of us with it. Its motor, its driving force, is the irrationality that is its life-blood and its poison.

The question for us is how to counter the system's irrationality with a vision of human development and interaction that is rational, sustainable and nurturing of humanity's abiding genius. How do we build a system that allows us to fully contribute to it and, by extension, to the well-being of all humanity?

Seven Technology Tasks for Our Movement

"A primary reason why millions have been able to mobilize so quickly is because they have the ability to use the open internet to communicate to the masses and organize a resistance."

-- Malkia Cyril (March, 2017)

The Internet is a classroom without walls, a library accessible without a card, a meeting room that requires no expense to reach and is open to all who seek to enter. Its power of access to so much information and communication is expanded by the Web's inclusion of you, and every other human being, as a source of information and collaboration. We not only learn what others think and know from the Web, we are free and even encouraged to add our own viewpoint, knowledge and experiences to that massive mix of information. By adding the hyper-link to this system, its developers have erased national boundaries, combated cultural exclusivism, battered racism and sexism, smashed into human isolation, gone a long way toward combating ignorance and expanded our ability to effectively write and communicate.

It's an essential tool in realizing the "rational society" we all seek but, to do that, we have to wrest control of the technology from corporations and governments...and that seems daunting.

The corporate and government control of the Internet's technology is pervasive and, in many ways, culturally unchallenged. Even people committed to revolution routinely use corporate technology without questioning the impact of that choice to communicate about the battle against the very corporations that are hosting those communications.

It feels like a Sisyphean struggle but that despair is self-deception. We can win this fight because, in the end, this control over our communications is based on the illusory power of their laws and their reliance on our usage of their systems.

As it has by profiting off the labor of workers, capitalism thrives off a system that is only viable because we, the great majority of the human race, use and accept it and don't really see any alternative to that situation.

In fact, nothing has changed in the Internet's fundamentals. It all continues to depend on the movement of data through packets of zeros and ones that we, through our keystrokes, control and that gives us a strategic opening, starting with challenging their "ownership".

And that's the first task that confronts us.

1 -- Challenge Ownership

"You can mass-produce hardware...(but) you cannot mass-produce the human mind."

Michio Kaku

The right to challenge corporate ownership of the Internet is based on an understanding of this massive network as a communication among people. Seeing the Internet as the massive act of collaboration that it is, the meshing of the human mind, makes wresting control of it from corporations not only advantageous but completely logical.

It's also a human right and the reluctance of world government convergences and bodies to resist declaring Internet control a human right is based on an understanding of it as just those machines and wires we've been talking about. Corporate control of the human mind is grotesque and impossible to fathom. If we believe that thinking collaboratively is a human right, so is the Internet.

To take on the political task of regaining our digital communications, however, it's necessary to take on the issue of how those communications are shared.

If our Internet uses phone lines and satellites as its "basic wiring", it can never be our Internet. In one of those cruel historical ironies life deals us, it would take a revolution to take back that basic infra-structure and liberate our communications but, to make a revolution, you need a liberated communications system that companies can't shut down. So you need control of the infra-structure in order to really control communications.

58

It's a two-step process. First, you need to understand that the Internet is a mind mesh and determine that its independence from destructive corporations is critical to our survival. Then, when you're convinced of that, you need to take on how that mind mesh can be done without that corporate strange-hold.

There are technologies that can address that. We can, in fact, build our own systems of connection that are completely independent of the corporations. The technology is here now and, with some innovation and organization, it's accessible to many communities. But that can't be done in one swoop; it must be part of a staged strategy.

That strategy could begin by building a localized Wi-Fi network using our own houses and computers to build a chain of access. That in place, an entire community would be on-line without significant cost and at the fastest speeds available. Such a system would give us a security we currently don't have and, most of all, implement a community-based model for Internet access. It would also help protect us from the kind of "access shut down" that would crush any opposition movement. There are already programs implementing this model.

A companion, or alternative, model for Wi-Fi access is one being tested in rural areas by progressive organizations like the Highlander Center. This is the purchase and placement of towers in high places (like the Tennessee mountains in which Highlander has its buildings). Using a system of Wi-Fi access points placed in strategic locations, you can "wire" a rural community for a few thousand dollars. The system does require bandwidth from a commercial company but so does everything else until the seizure of power. We'll get to that in a moment.

We can then implement a "peer to peer" network that turns everyone's computer into a server that is part of a network of servers that, together, can handle any amount of data commercial services can. This is the absolute alternative to corporate and centralized Internet storage and data distribution. It resolves the issue of Net Neutrality -- the right to transmit and receive data at the same speed as everyone else without paying extra -- since we are now transmitting our own data. It also enhances privacy and security.

These are steps. They are not independence. To achieve full independence, we would have to either develop alternatives to the current structures of communications power (telecoms and other communications companies) or take over the ones that exist.

There is no strategy for doing that and to describe of one now would be a waste of time. We don't have a movement capable of completing such a task and so what would be the basis for that strategy? In fact, the construction of a strategy goes hand-in-hand with building movements in a developmental dialectic some of us are only now starting to think about.

The truth is that the complete independence of the Internet can only be a product of revolutionary change and we need a movement conscious of that technology task to make that revolution happen. The construction of an independent WiFi network and a "peer to peer" network would provide our movement with alternative forms of communication that would broaden that movement, give it experience in doing this kind of alternative work and provide an increasing better focused vision of what to do about technology. In short, it keeps us connected, broadens the population of others who are connected and provides a basis for and practice in these strategy discussions.

2 -- Challenge Surveillance and Declare Full Privacy a Human Right

"I always feel like somebody's watching me...and I have no privacy"

Rockwell, "Somebody's Watching Me"

Given all we've said about surveillance, we can keep this short and to the point: our movement must prioritize an abolitionist position on the capture and use of people's data by corporations and governments.

That capture violates the fundamental principles of the Constitution of this country and every aspect of democratic functioning. Given the use of this surveillance as a primary weapon in the arsenal of repression, we have to free ourselves from it or we will not be able to change this world.

Certainly the struggle for stronger legislation against surveillance abuse is a worthwhile effort and should be supported by everyone. But it's not enough. The world's governments are in constant collusion and the Internet's borderless communications make privacy protection laws in one country almost meaningless. Fighting for more legal protection is only a step toward the more significant goal: establishing the right to full privacy on the Internet as a fundamental human right.

Much of your communications has no value if it isn't private. No government has the right to disrupt that privacy with surveillance and no corporation has the right to capture information that you transfer or save with the expectation of privacy. The expectation of privacy is not a negotiable concept; it's an absolute right. If the Internet is to be of any real use for us in changing this world, we have to prioritize the privacy we should be confident of when we use it.

That should be a universally expressed demand within all movements for change all over the world.

3 -- Change the Development Culture

"Movement technologists and movement activists must develop a deeper collaboration emanating from an understanding that we are both essential parts of building powerful movements."

Movement Technologist Statement [12]

Information technology is motored by computers that are guided by the "programs" we discussed before -- written, tested and continuously developed by developers – and that interaction is managed by other people commonly known as "administrators". Technologists play both roles.

The roles these "technologists" play are central to information technology and to the effort to re-define our relationship to software and its relationship to revolutionary change.

The "development culture" in capitalist society views all developmental work as a product and views those who use the hardware and software developed as consumers. That's the ideological underpinning of the commercialization of information technology. This usually requires little discussion within our movements. Movement activists are very clear about how distorted and obscene this frame is.

Yet we continue to accept it in the conduct of our movement work.

Historically, technologists and the role of technology have been excluded from wider political discussions in our movements. Technologists are still viewed, by most movement activists, as sources of expertise whose main role is to provide the movement with its technology tools.

In fact, that role is real; it's what we do. But that relationship blocks our movements from gaining the kind of control over our technology development work that is a first step toward the full democratic control of the technology itself.

If someone else is producing your software, without your collaboration, you simply have no control over it. As a result, not only does it not respond to the technology needs you have as an organizer but it doesn't move us forward in the struggle over technology we have to engage in.

At the same time, this separation hinders the political development of technologists, sharply separating them from the day to day struggles people are waging and that our movements are leading. That robs our movements of these people's thinking and commitment while making their work less reflective of the needs of those struggles.

This is the principal issue taken up by the Radical Connections Network, the network formed by movement technologists and technology users in the United States that issued a statement in 2018 partly quoted here:

"Historically, technologists and the role of technology have been excluded from wider political discussions in our movements. There is one thing needed to start developing a deeper relationship between technologists and technology and the broader movement: movement technologists and movement activists must develop a deeper collaboration emanating from an understanding that we are both essential parts of building powerful movements.

"Effectively, our movement should consider its technologists as the primary sources of information about technology, moving away from the toxic dependence much of our movement has on corporate technologists and tech companies. At the same time, movement technologists and tech organizers should be invited to lead the way on new tech practices and platforms for our movements, moving away from the all too prevalent tendency of developing software and other tech solutions without direct input from non-techie movement activists and organizers." [13]

Can it be stated any more clearly?

4 -- Claim the Right to Our Story

"Stories are the threads of our lives and weave together to form the fabric of human cultures. A story can inform or deceive, enlighten or entertain, or all of the above at once. We live in a world shaped by stories."

Patrick Reinsborough [14]

For our entire existence as a species, our communications with each other has been based on the stories of our lives. Our life experience is really the only tool we have to understanding our world and our relationship to it and human beings sharing that experience is our only means of survival.

The various stages of human development mirror the expansion of the systems through which we tell our stories, from the primitive societies in which those stories were shared among a small group of people to the periods of feudal production in which those communities of story-telling were expanded to towns and villages to the various stages of capitalism in which larger and larger groups of people living in cities and nations shared their knowledge with each other in order to produce. At each stage, we've invented forms of communication that group together larger and larger numbers of people.

The rulers of these societies have developed methods of framing those communications, designing those stories and communicating them through myths, religions, fables and later through mass media...all the time seeking to imprint on those collaborative stories an analysis that favors their continuing control.

The history of humanity is, in part, a battle over who will tell our story. For example, the African sugar cane workers enslaved in my native Puerto Rico told their stories through songs and dances (known as bomba) which they celebrated in surreptitious events out of the eye of the plantation owners who would punish them when discovered. That effort to crush indigenous communications forms was a typical reality of slavery and the insistence in practicing them was an act of resistance vital to the culture and communications of the enslaved people. The push to "convert" enslaved people to western religions and "teach" them "civilized" forms of communication contended with the people's relentless desire to "frame their own stories".

So it's been for all human history. We seek to frame our story and the powerful seek to repress us when we do it.

The contest over who will frame your story has become most intense with the Internet. Not only does this network facilitate communications between larger groups of people than ever before but it shapes their collective expectations, their views on culture, morals and success, their understanding of their lives and, perhaps most importantly, their participation in the lives of others. Because the Internet is so massive and provides so much access to people, it provides us more opportunity than ever before to tell our own stories.

Of course, the ruling class isn't asleep; it tries at every turn to take control of how we tell those stories, limit how much we contribute to that telling and frame the reactions of others to our stories.

The question in contention is "how is our story told and who tells it"? In the end, that question is the one that defines human behavior, culture and society – now and in our future.

For most of human history, that story of our daily lives has been assigned to story-tellers who were part of the community they served and their stories were designed to mirror our lives and be consumed by those who are living them. They were a mirror of sorts, reflecting the essence of the reality of our lives while interpreting that reality in ways that may, or may not, resonate with us.

In some cases, they framed that interpretation through the lens of the oppressive ideology and, drilled by that interpretation, people came to accept it. In others, as with indigenous cultures, they represented an act of rebellion against the oppressive story: often using tales, song and dance to filter truths about our past and our present into the minds of people in that community.

In contemporary society, those story-tellers are media workers: journalists and the people who circulate their journalistic work. But there is huge problem with that, emanating from the development of the society we now live in.

Journalists are privileged people extensively trained to cover, not our lives, but the decisions and actions of powerful people and institutions that affect our lives.

When I worked as a daily newspaper journalist I was frequently told by editors that the key to any news story is to get "the official reaction" or "the official line" and then, if it fits, some response from "regular people". The only exception would be a feature on those regular people for which we would dip into their lives for a couple of hours, get some answers to questions we framed without knowing whether they had any real relevance to these people's lives, and then fit them into a narrative that our readers "would understand".

The people who weren't "our readers" (a fictitious and arbitrary construct at best) never spoke except through me and my colleagues.

That media practice has resulted in a huge devaluation of people's ability to tell you about their lives and, by extension, a devaluation of their own opinions of those lives' importance. Reality is filtered through the perceptions and, by extension, life-experiences of relatively privileged people who are, more often than not, completely removed from the people whose lives they are reporting on.

You can't capture a person's life experience in a quote or a one-paragraph description based on a quick and superficial look. Even the most accomplished and committed journalists -- and there are many of those -- distort reality as much as they may want to accurately portray it.

The Internet smashed through that problem explosively by expanding on the other "frame": people telling their own stories. Now anyone can tell their story and share it with anyone else who wants to read it. Stories carry lessons and those lessons frequently include ideas about and examples of how people survive. In many cases, they detail actual struggles by communities and movements. Some draw analysis from those struggles.

The sharing of all this information, the meshing of the mind, is a huge step forward in humanity's struggle for survival as it enables the functioning of the human mind. The fantastic growth of the Internet has broadened and deepened that story sharing.

There is a resistance to that, however, and it takes several forms.

The stunning resistance, by the capitalist class and its government, to what is called "Net Neutrality" is a case in point. The issue is most frequently framed as a kind of consumer right: you have a right to receive data at the same speed and ease of access from all sources (big companies to small websites) and telecoms and other access companies have no right to provide quicker connection speeds to big companies that pay them more than, say, an organization that has one website and can't pay the higher price. Your connection speed may vary depending on the price you pay for the connection but everything you connect with needs to connect at the same speed.

The debate has raged and has proven subject to the politics of the particular Federal Communication Commission majority that is in power; a right-wing majority has almost destroyed Net Neutrality as I write this although that struggle continues.

This "consumer rights" battle is an important fight to engage but this is not a consumer issue alone. Net Neutrality is about your ability to communicate with other people freely and efficiently so you can share your stories. Most of the people you are going to share with can't pay a higher rate for their connection so their data is going to flow on a much slower track and, in a very short time, it will be much less possible to efficiently get that data.

In fact, given that most people on the Internet will be attracted to a faster and more robust download, your information will become less and less popular and visible and may actually disappear. The sharing of our stories, the essence of the Internet's collaborative power, is directly threatened, as is the flow of information about issues and rallies and campaigns and incidents: the stuff of movement communications.

The attack on Net Neutrality is a form of repression no less dangerous and destructive than throwing activists in jail or seizing their means of communication.

The same push towards restriction takes other forms, sometimes in patterns that may seem to go in the opposite direction.

Social media, the most expansively popular Internet protocol of our time, has the huge benefit of bringing more people together than ever before but it restricts what they can say to each other.

Facebook is the most powerful social media service and it serves as an excellent example. As powerful and robust-seeming as it is in our lives, it is actually constraining and divesting of our power as communicators. In order to use it, you have to use it the way they want you to and that's not a whole lot of "using". Certainly, there is a comfort in having one's options limited, being able to use something without learning anything about it or making many choices about how you use it but that alluring convenience is a poisoned apple.

In order for Facebook to most efficiently gather information about you, which it will share with anyone who pays and everyone who governs, it has to impose a uniformity of presentation and routine that can never be changed. A Facebook page by you looks and acts exactly the same as a Facebook page by everyone else. If you want to try to alter what's presented and how, you can't.

One of the benefits and strengths of the Web is your ability to design and organize your presentations on a website that can easily be made unique: showing what you want and hiding what you don't, protecting contact with others through web forms and discussion boards that let people "hide" their real identities, controlling what you share with the world and presenting it in a way that actually enhances its meaning. With a website, you can exercise the power over your message about yourself, the profile the world has of you, the presentation of the life experience you want to share with the rest of humanity.

That can't happen with Facebook or Twitter or any of the other major "social media" platforms which display information about you while restraining your ability to contribute information and thinking about the rest of the world.

With Twitter, for example, you have a limited number of characters to make your statement. How much thinking can you communicate in 280 characters? Twitter feels like a room in which a large number of people are shouting single sentences — a lot of noise, even a few ideas but mainly just individualized statements bereft of context, reference to supporting information or the need to exchange perspectives.

Facebook is equally constraining with so many one-sentence statements that writing anything longer seems strange and even rude.

Just how these limits work on people's communications is highlighted by a startling fact. When Twitter expanded the number of characters you can post from 140 to 280 characters in late 2017, the expansion had almost no impact. Data from the company shows that, in 2018, only one percent of the Tweets posted hit the 280 limit and only 12 percent exceeded 140. People's use of Twitter had trained them to write short, self-limited posts and the limit expansion changed nothing.

The incremental "take-over" of the Internet by these programs oppresses people, particularly young people, by repressing their communications and, by repeated practice, their thinking. It strips the very powers the Web has given us.

It also puts in the hands of corporations and the governments with which they collaborate the power over vital communications because it's easier to shut down a Facebook page or a Twitter feed than a website.

To shut down a website, the government must obtain legal authority through a hearing or some clause in the law and then impose that on the hosting provider. That requires the collaboration of the provider, not always guaranteed. In most cases, they can do it but it's not easy or quick and speed is often the essential in a repressive strike. They must shut the site owners down before the owners announce that they are under attack because, if they get that announcement out, the repressors face serious opposition in the streets or courts or on-line.

Because Facebook and Twitter are, effectively, a contained website controlled by the company, they can shut a page down in a few seconds and the user can do nothing to prevent that. Complaints by people shut down by Facebook have become daily occurrences made more frantic because the Facebook staff doesn't have to tell you precisely why you were shut down and, in most cases, refuses to.

The balance is between, on one side, the threat of easier repression and very real constraint over communications and, on the other, the mass popularity and cultural ubiquity of these communications forms.

Our response has to be to emphasize our exercise of those platforms' benefits. No one can seriously argue that we should discourage the use of social media. It's just too large and popular and is now too deeply ingrained into the culture of young people world-wide. Yet it's possible to advocate, starting with the way our own movements use these platforms, for a more targeted approach.

One example: Media Justice, possibly the most important movement organization working around information sharing today, effectively uses social media for calls to action or to identify an important news story or development. That delivers a quick and effective message. Yet, the organization has a robust and extremely active web presence updated daily with news, opinion and analysis of events – more expansive and richer content – to which it constantly points in its social media posts.

The challenge to social media isn't a challenge to its existence or even its use. Those, at this point, are foregone conclusions and a strong case can be made that social media has a powerful and productive function for any community and any movement. The point is to situate it in the arsenal of on-line communications based on its strengths and its substantial limitations. It's the initial message, the siren call about events and the seductive invitation to engage and go deeper: a kind of neon sign on the door of each of our lives.

Using it properly, we enhance our engagement with the rest of humanity. Depending on it solely, we become entrapped in its constraints and risk putting everything into a silo that could, at the whim of a corporation or government, be closed.

That's the choice we're going to need to make if we seek to democratize our voices by telling our stories.

5-- Challenge the Existence of Nations

"There are no nations! There is only humanity. And if we don't come to understand that right soon, there will be no nations, because there will be no humanity."

--Isaac Asimov

Among the most important fictions our culture has absorbed is the bizarre concept of separate nations: people divided into territories with clear and unassailable borders that are protected by people with guns and governed by people with political power.

So deep-seated is this mythology that even the most politically critical among us hardly question it. We organize within those borders, travel within them, think and analyze restricted by their boundaries. We hardly ever think of a political strategy that crosses those borders let alone encompasses people in other, more remote, parts of the world.

National borders are a major impediment to world social change because they are the most potent disrupter of world-wide collaboration.

Not only are we restricted to political activity within our borders but we are routinely cast into a false competition with people from other countries and, more and more, into actual conflict with them. You can't have wars without borders nor can you have the demonization of people from neighboring areas who seek, for whatever reason, to cross them.

The "immigration issue" that is currently drawing people's attention in the United States (and in many European countries) and producing a spasm of nasty, hate-filled mass reaction is all the more frustrating because it's about people crossing artificial divisions. Is there any sane and logical explanation for the "border" with Mexico?

The logic of the border melts away when our history as a human race is examined. Countries have been fabricated by mashing together groups of people whose cultures may have little in common, whose languages are often vastly different and whose pasts are driven by regional aspirations that aren't common and sometimes conflict. In fact, many of the world's nations are born of a genocidal elimination of the people who were there first. That's certainly the case with the United States that couldn't qualify as a "nation" based on any historically defensible definition of the term.

The concept doesn't make sense given our past and is absurdly contradictory with our present. If we make prominent the thing we actually do as humans, produce so we can survive, the world we live in isn't divided by nations at all.

Production is performed world-wide with people making products that others in their country may well never use or even see, in languages that they probably don't understand and packaging that, in their culture, makes no sense at all. Those products are consumed by people living in a society that doesn't make them, like most of what we consume in the United States. We survive on the labor of the entire human race and there are no borders to that labor or the consumption of its products.

We also share with all others on earth the suffering our condition creates. Climate change and its ravaging destruction doesn't respect borders. Hunger and deprivation aren't divided by walls. The reality of women in one society mirrors the punishing situation of women world-wide. The brutalization of people based on their ethnicity, race or background goes on every place where humans live. We're united by our condition.

That makes for a whole lot of potential power and logically the response to our shared distress at our oppression would be to ban together and make the changes that are needed. That's where the national borders flex their cultural muscle by keeping us divided in our consciousness as well as actually divided in our knowledge of each other's situations.

So what does the Internet have to do with national borders?

The routine of on-line communications respects no borders and easily, and quickly, crosses them all the time.

We visit websites hosted in countries it would take a day to fly to. We read email from people in places we probably will never see. We share thinking, information and strategy with others in political meetings attended by people in various parts of the world...without any travel. The experiences we share across this world without borders unfold in vastly different conditions involving people whose cultures sculpt experiences that are dramatically different from ours. We all learn from those experiences and, in the process, our thinking is influenced by an ever-greater perception of what life is and what it can be.

This makes critical the battle to maintain a free Internet in every part of the world. The repression of communications in countries we may know nothing about is a crippling blow to our development and ability to survive.

Many of the organizations that carry on the struggle for a fully accessible Internet, like the Association for Progressive Communications, focus on the specific laws and policies of the governments that rule affected people. A day doesn't go by when APC isn't raising an alarm about the attempt by a government in Africa, Asia or Latin America to bottle up Internet communications or chop off part of its functionality. That is one of APC's most important roles, deserving of all our support.

That, however, isn't enough. The banner that has to be flown is that the Internet makes clear the obscenity of the very concept of a border and contradicts the viability and usefulness of countries and governments.

The movement for Internet democracy must reject all borders and we can start by communicating with greater international inclusivity, consciously seeking to make connections with people in other countries confronting similar issues and doing similar work on them. We can insist that Internet rights be recognized in every country no matter its laws and that such evils as censorship or restricted access represent human rights violations.

Not only do those actions enrich our political work and its potential outcome, but they are first steps in abolishing borders in all human activity and the act of rejecting borders in all human activity is a major step toward the revolutionary change that is essential to human survival.

For a human race seeking to survive, there can be no recognized borders.

6 -- Challenge the Limits to the Imagination

"Imagination is not only the uniquely human capacity to envision that which is not, and therefore the fount of all invention and innovation. In its arguably most transformative and revelatory capacity, it is the power that enables us to empathize with humans whose experiences we have never shared."

J. K. Rowling

Within the Left of the United States, there is a prevailing malaise about winning. It's palpable in meetings or conferences, an ever-present block to thinking about long-range change. It's not that people don't want to win or that they're not committed to it. It's that the goal of that victory isn't clear and the countering force of oppression so powerful and encompassing that victory is virtually impossible to perceive.

The rise in the 1990s of what is now called the "Social Justice Movement" did little to clarify things. That fact doesn't diminish its validity or importance. As the most important political development of the last two decades, the movement has contributed mightily to the struggle for human survival and resistance. But its very essence, formed in the conditions that created it, contains a built-in limitation.

The movement arose out of the ashes of the revolutionary movements of the 60s and 70s, in response to a combination of three related developments:

* The continuing social crisis and the brutal erasure of many of the gains won by the previous movements of struggle.

Every major gain we've made, including those won during the New Deal, the labor movement's victories, the Civil Rights struggles, the victories of the women's movement and the massive reform of education -- all it -- has been either reduced or completely taken back from us.

* The radically altering entry of people of color and women into the ranks of universities, spawning generation after generation of increasingly educated young people from the very communities that were most impacted by the destruction of the social gains.

This provided these young leaders with a massive amount of data and analysis coupled with consciously developed analytical and communications skills. Today the Social Justice Movement is essentially led by people of color and women (very often women of color). That development transformed a left-wing movement that, during the 60s and 70s, marginalized these same people and frequently kept them out of mass leadership.

These weren't "reformists" in the traditional sense. Reform work for these new leaders was strategically targeted and seen as a means to a socially transformative end and their political vision dug not only into the traditional areas of observed oppression but to new layers of social struggle including areas of gender identification and forms of leadership development. A movement that had, twenty years before, thrust activists into action like a kid thrown into water to learn to swim, now engaged in complex and protracted programs of training and development.

Yet that very concentration on building the skills and analytical abilities of leaders also robbed the movement of the time and even interest in the kind of movement-wide conversation needed to build a revolutionary strategy in the wake of the collapsed proletarian strategy. Doing that would have required a special effort and the resources for that kind of effort weren't there because...

* The entry of foundations into the quest for social change.

Previous left leaders lived and worked in a more affordable society with a greater prospect of employment that could provide some financial stability while giving them time to do their work. Some even had families and parents who were a source of financial assistance.

But this new leadership came from a different demographic and these new leaders had few such financial prospects. In many cases, their families expected them to provide some relief from persistent poverty: the reason many of them started college education in the first place. This different social situation and demographic meant that they needed to be paid.

73

The collapse of the mass left-wing movement of the 70s coincided with the introduction of foundations into the social struggle. Fed by capitalism's wealth, these foundation began funding struggle.

The problems inherent in depending on foundation money to support struggles against the very system that creates the wealth foundations distribute is exhaustively covered in the now-famous book "The Revolution Will Not Be Funded".

In that still very relevant book, activists and scholars analyze and illustrate (often recounting their own experiences) how foundation funding creates a crippling dependency for social movements and that the dependency more often than not steers them away from revolutionary goals.

By becoming the arbiters of the political choices funded organizations make, the foundations dull the politics of those movements and pressure them into reformism. They also create a hierarchy among movement organizations that is very often artificial: not based on the support or work impact these organizations have but on the skills of their fund-raisers and the relationships leaders of these organizations can build with foundation officers.

Again, some balance is needed. Much of the foundation-supported work in this country is extremely important and impactful. Reformist work often changes people lives for the better; in many cases, it saves those lives. But that work will always stay within the boundaries of reform if funded work is the only work an organizations does.

In fact, if the organization takes on other "non-funded" work that is specifically revolutionary, it risks losing that funding. In any case, it can result in a fatal separation between the work people are doing and its impact because the measure of success is not necessarily how much popular impact the work has but how successfully an organization can fund-raise for it.

Social justice leaders have become expert at framing their work to make it seem less dangerous to this system than it actually is: a prerequisite for funding in many cases. Still, the work must be reported on to assure continued funding and those reports are scrutinized against a set of "required outcomes" that seldom if ever involve raising mass consciousness or sharpening the thinking around long-range strategy.

The bottom line is that no foundation is ever going to fund a revolution.

Those three pressures have created in the left of this country an inability to see beyond the work that is being done. With the erosion of previously won social gains, the survival-threatening crisis felt daily by communities in which this movement works makes immediate response a priority and immediate results a necessity. That pressure is felt personally by this new leadership because they often come out of those battered communities and, in many cases, still live in them, experiencing the crisis first-hand.

Finally, the pressure of performance and the rigorous demands foundations make to report on the outcomes of their funding investments means that intellectual capital is spent, not on a long-term strategic vision, but on figuring out how to meet the outcomes and renew funding.

This is a crisis of the concrete. Strategic thinking in this movement is seldom allowed to go beyond what is immediately achievable and the collective mind of this movement is blocked, by time and focus, from engaging the imagination in the mapping of a strategy.

If we learned anything from our Technology and Revolution project, it's how the conception of what's possible is imprisoned by what is currently visible. We find it difficult to think beyond what we see and the oppression we see is most easily countered by reforming what is being experienced.

The problem with the kind of reform work that is the specialty of the Social Justice movement isn't that it doesn't improve things. Every major reform in this country's history -- from those of the New Deal to the dramatic changes brought on by the democracy movements (both the Civil Rights movement and the women's movement) to the thousands of "smaller" battles over rights and privileges -- has produced fundamental and often vital improvements in people's lives.

But, in all cases, reformism is pressuring the ruling class and its government to step back off oppression and provide us some breathing room. If things remain in that realm they can be taken away and, as is happening today, will be if that's what capitalism needs to happen.

The challenge we face as a movement is to imagine the world we want and that world, of necessity, requires a change in who is running it. It requires the end to capitalism as our economic system, a democratization of decision-making which is impossible with the current government structure, a radical shift in our relationship as a species with the Earth we inhabit and a radical change in how we view work, productivity and social contribution.

For most of my life, such a world was seen as feasible only after a period of development in a revolutionary society. The human race had to take charge and then, in a protracted process, it could build toward realizing our social dreams, transforming the imagined into reality. We didn't have the physical and technological capability to do that immediately.

Now, for the first time in my life, we do.

Everything you can imagine in this world and dream about is attainable because we now have the technology to do it. Information technology frees us from constraint. You can feed the entire world, build shelter for it, arrange it so that it's protected from weather-related destruction, elevate the true interests and talents of everyone to a place of productivity, radically re-design the structures and methods of democratic decision-making...all with the technology we now have in place.

Those changes, democratically planned by all humanity using that technology and its communications, can profoundly change human culture, altering the view people have of each other and the relationships we enjoy.

The challenge is how we free that imagination and that entails engaging in an effort to democratize the technology and wrest control of it from this small group of destructive rulers, jaded by their self-centered worldview. In that struggle for the democratization of technology, our collective mind will increasingly exercise our imagination and absorb its lessons because, for the first time in history, we have a technology that can let us exercise that collective mind.

7 -- Change How We Live in the New Climate

"Treat the earth well: it was not given to you by your parents, it was loaned to you by your children."

Tasunke Witko (Crazy Horse) -- Oglala leader and warrior chief

All thinking about our lives and our societies is enveloped in a new and frightening reality: we are facing extinction.

76

While scientists speculate about timing and politicians debate urgency, there can be no rational disagreement about how radical changes in the world's climate, weather conditions, water levels, food production and species existence loom as the central threat to our lives. No issue can overshadow that fact and all struggles against our oppression must take it, and the time it gives us, into primary consideration. Doing what we currently do in relationship to our Earth and living as we currently live on it are simply not sustainable. We will be wiped out as a species if we don't radically change our lives.

Today, an estimated eleven percent of the world's population is currently vulnerable to the droughts, floods, heat waves, extreme weather events and sea-level rise that are the products of climate change. As we gallop forward with the abuses of the Earth that cause all that, that percentage is guaranteed to rise. At its current levels, it's already a huge problem because the human inter-relationship in food production and the population movement or "migration" of desperate people seeking to survive is causing an imbalance in resource that our current unbalanced system is unable to cope with.

The discussion about putting a brake on climate change through the control of carbon and greenhouse gas emissions and other pollutants linked to production is finally assuming a front seat in the political debates of our time. The priority for the human race is to stop it right now and if that's impossible given the systems under which we live, the priority is to change those systems.

Yet halting the rise in temperatures, the main culprit in this extermination threat, is not the complete solution to the problem. The damage that has been done is not reversible. The world's temperature is not going to be lowered. Even if we stop the environmental abuse now, this nightmare of flooding, crop disappearance, animal elimination and punishing heat is what we're going to be living with and adjusting to from now on.

What's required in this newly challenging situation is a major change to how we live and how our living places are structured. We need to plan how over 7 billion human beings can live cooperatively making all competition a social fossil. Our communities must be redesigned to respect the power of flooding waters. Homes must be designed to resist the punishing impact of unprecedented storms. Food production must be re-evaluated to stress a more economical use of space while availing ourselves of environments that can be protected from unpredictable weather. Food distribution has to become coordinated among all societies based on need and potential. Our diets have to shift. Our travel must become less expansive and more secure. The production of our necessities (from tools to furniture) has to be rethought and, as much as possible, more localized.

We have to build community living places that are planned for security and, most of all, sustainability.

As daunting as that might seem, information technology makes it possible. The collaborative thinking required to do this planning and changing is within grasp because we can communicate with each other instantly with no distance restrictions. Anything you can say to someone in person, any study or information, any diagram or effectiveness test, any observation of their circumstances and demonstration of yours...it can all happen in a moment.

With information technology, we can produce any non-organic thing we need. What you sleep on, sit on or walk on can be produced locally with computers and designs shared world-wide. Nothing has to come from the other side of the world and, with the knowledge we can share, things can be made more economically and sustainably than ever before while employment can be designed and expanded based on local needs.

The point, however, isn't how information technology can make possible this new sustainable society but that these necessary changes are impossible without information technology. Given the time pressures, the huge differences in the situation in various parts of the world and the enormous developmental imbalance among living places that capitalism has created, the coordination and collaborative planning needed to save our species isn't possible without immediate, broad-based communications and sharing of ideas. There is no way we can rely on discussions at conferences or jointly developed resolutions of intent. We have no time to waste; we can ignore no corner of the world.

The survival of the human race is only possible through the intellectual collaboration, the meshing of the minds, that information technology alone makes possible. Concentrating on that must be a priority.

A Program of Action

"At this crucial time in our lives, when everything is so desperate, when every day is a matter of survival, I don't think you can help but be involved."

Nina Simone

In those Technology and Revolution sessions, between 2017 and 2019, those 1,500 activists in 25 the convergences in two dozen cities of the United States and Mexico met to talk about the intersection between technology and fundamental social change and to try and hammer out a short program of action of our movements. Each convergence considered the consensus from previous convergences and tried to amend or enhance them. The convergences, under the title "Technology and Revolution", continue to this day.

In the Spring of 2019, the coordinators of this convergence program took stock of how far we had come and what points of unity had come out of the convergences up to that point.

This is a program of action around technology, designed to move us forward in our struggle for a new society, summarizing the hundreds of hours of conversation among those activists. It will probably continue to change, as should all that we decide on, but the consensus program is the best expression of what our movement thinks about what we should do.

• Provide everyone with full, FREE, high-speed, equitable, access to the Internet, independent of content.

• Build an internet that is democratic, community-centered and governed, open, decentralized, and free of corporate pressure and monopolies.

• Build a political technology campaign to oppose, restrain, and ultimately eliminate surveillance.

• Move technology to prioritize sustainability, community thriving, climate justice and "many worlds are possible".

• Seek out, build, and embrace the potential of digital technologies to protect and advance our movements.

- Improve and deepen the collaboration and mutual education between movement technologists and other movement activists and organizations.

- Foster political consciousness about the centralization of technology in movement work and the urgency of revolutionary movement-based technology.

- Expand the technology conversation beyond settler/colonial technology and thinking to be culturally relevant, inter-sectional and grounded in political education and historical context.

Summing Up: Freeing the Mind

"Remember you have within you the strength, the patience and the passion to reach for the stars to change the world."

Harriet Tubman

What if we could implement that 8 point program? What would be possible? To answer that question, we have to re-frame how we are seeing technology.

You take a trip in a car and it fills you with experiences and memories. Those experiences and memories, however, are very seldom about how the car operated: the role of the carburetor, the firing of the spark plugs, the traction of the tires, the consistency of the steering mechanism. You don't analyze the internal combustion or the automatic gear shifts. Unless the car broke down, you give its performance and the technology on which its based no attention. You're conscious of how important all that is, particularly if it stops functioning well, but that's not what the experience is about.

Rather, you'll remember the sensory experiences like what you saw or smelled or felt. You'll remember human interactions you had along the way. You'll remember the feeling you had when you arrived at a destination.

You take the functional elements of the technology for granted because, in the end, they are less important than the human experience that technology made possible.

By convincing us that the Internet is basically the technology that runs it, the system which makes functioning possible in so much of our lives acts to restrain our imagining about those lives. It sticks us into a prison of the current, crippling our ability to see it beyond how it currently works. We find it difficult to imagine communications without the current hook-up of computers, without the systems of connectivity and, by extension, without the corporations and powers that control all that and distribute it to us based on their needs and interests.

Stop thinking that way for a moment! Try to envision this digital communication, this Internet, without the wiring and programming. It's not easy. Karl Marx, in his introduction to the second edition of the first volume of Capital (the only volume published during his life) reflected on the difficulty of getting to the core of systems. He warned readers of that volume that they would be challenged trying to understand the first sections (where he and Engels explore the Law of Value). To get to the essence of systems, a scientist must get beyond the obvious and the surface, he wrote, and laboratory scientists have equipment (like microscopes or chemicals) to do just that. But, Marx wrote, "in the analysis of economic forms...neither microscopes nor chemical reagents are of use. The force of abstraction must replace both."

We aren't taught to think abstractly because our culture teaches us that experience is best understood by what we can see. But you can't view the future through what you can see. To view its future, our species has always relied on its uncanny ability to imagine, to reform what is there, to erase what is less important. That is what we have to do when we look at the potential of information technology and digital communications. The Internet, for those of us looking to the future, isn't what makes it work but what it does and can do for those of us who use it.

See it as the people who use it, understand its power as the information they share, see its potential as the limitless sharing of the mind.

To do that, we need not go into the mystical experiences some of our ancestors had. We can envision this limitless sharing as something that is actually possible right now. We have the technology in hand to free our communications and our collaborative mind.

What if we eliminated those large wires that connect us to the network? What if we didn't have to rely on some corporation to connect us or to service our community at an exorbitant price because wiring that community (maybe a rural community) isn't cost-effective? What if all connection was handle by large towers in each neighborhood?

What if, after freeing ourselves of those trunk wires and the companies that own them, we turned our own computers into servers? The computer on your desk then serves, not only to handle your data but to share the data you want with other computers across a giant, linked network? What if a part of your computer serves as a storage and data transfer unit for other people's data: people you don't need to make an agreement with or, for that matter, people you don't even know?

What if this giant network of intellectual collaboration were to take collaboration a giant step forward, turning it into a massive hive of interacting computers whose task is not only to do what you want but to do what the network needs?

All of that is possible with the technology we have in hand right now and we can take it a powerful step further.

What if we take data technology beyond the physical bounds of that box with its wires and circuits and storage devices. What if our entire home were to become computerized allowing us to store ideas and messages and information at any moment by merely speaking it or reading it or feeding it into the system? What if we could connect that home system to the systems of homes all over the world?

What if we were to answer the call Tatanka Iyotake makes to us, a call we can finally answer using this technology: that we can collaborate quickly and freely and efficiently and completely to make that future our children need? What if the definition of the Internet isn't the Internet itself but the meshing of the minds of the human race it makes possible? That is what we need to preserve; that's what we need to expand.

What if every decision you make from the moment you're awake is informed by the experience and knowledge of the rest of the human race...and that decision adds to that knowledge and experience and informs the decisions of everyone else? A thought you have now influences the thinking of the rest of the world. An activity you engage in moves everyone else to consider whether that activity moves their lives forward. A demonstration or meeting or conversation you share with audio or video transfer becomes a model for millions of others. A discussion about our world, our challenges and our strategies for meeting those challenges can now involve people world-wide. Borders are meaningless. Cultures become backdrops enriching and helping us apply the experiences and struggles of people all over this small and struggling planet.

The way indigenous organizers fighting forest extinction in Brazil or peasant activists in China confront particular problems of unbalanced distribution become new ideas for how we, in the United States, can confront similar issues. Moreover, because those problems are immediately known to us, we can respond in ways that provide untapped and powerful solidarity...and we can expect the same from those people.

83

The lack of developmental balance, a product of illogical trade systems under capitalism, is no longer a problem because the people who actually produce can collaborate on that production and distribution. The frenzied destruction of our environment by a suicidal economic system can be halted because we each understand how violent that destruction is in many parts of the world. War has no logical purpose. Denying or minimizing the impact of climate change becomes ludicrous when we can see first-hand how it's destroying lives and communities world-wide.

Hunger is illogical. War is nonsensical. Division is unfathomable.

As enormous as the task may seem, as daunting as the challenges appear, we can do all this. We have developed the technology that makes possible our response to this moment of choice between nightmarish destruction and the realization of the world we, and those who come after us, deserve. In answering the call to "put our minds together", we can join our hands, defeat the monster and enjoy the goodies from the Yum Yum Tree.

Acknowledgments

About twenty years ago, I was in Toronto, Canada visiting a bookstore with a friend when I came upon copies of one of my books about Puerto Rico: Doña Licha's Island. As I handled the copy, a young woman standing next to me recommended it and I told her I was the author. She said she loved it and I asked her what was her favorite section.

"The introduction," she said. "Where you thank everybody you've ever known. I though that was so cool."

I remember the exchange as proof to myself that this idea of meshing minds, so important to the current book, was in my head for a long time. So I'll start by thanking everyone I've ever heard or spoken to or read for giving me all the ideas I've set forth here. I mean that. Seriously. I've never had an original idea in my life.

Specifically, I'll thank my work partner, comrade and best friend Jamie McClelland for his outstanding suggestions and remarks on the book, particularly the technology section since he's the best technologist I've ever met.

I want to thank my sons: Karim Abel López Arrastía for his careful reading and many critical ideas about presentation and Lucas Ernesto López Arrastía for the hours he's spent with me talking about and exploring social analysis and theory, which he knows so much more about than I.

Finally, I thank my life partner of over forty years, my wife Maritza Arrastía -- revolutionary, writer, editor and teacher -- for her careful edits of my writing, not a very easy task...and for making my life worth living, even harder and more precious.

And thanks to you for reading this book. I value your ideas, criticisms and questions. Write me. I'm alfredo@mayfirst.org

Alfredo Lopez - October, 2019

About the Author

Revolutionary activist Alfredo López has been involved in social struggles for over a half century. He has helped organize major demonstrations, particularly as a leader of the Puerto Rican Socialist Party, and has belonged to several political organizations including May First Movement Technology, an organization he helped found and in whose leadership he currently serves. He's written six books. He lives in Brooklyn with his spouse, Maritza Arrastía.

Alfredo can be reached at <u>alfredo@mayfirst.org</u>

1 - The Fetishism of Technology: Cause and Consequences, David Harvey
https://pdfs.semanticscholar.org/f8b9/af432945832dfcab4c0e30536706c7e66c51.pdf)

2 - Internet is all about collaboration, Olaf Kolkman
https://www.internet.nl/article/internet-draait-om-
samenwerking/

3 - Capitalism Versus the Environment, Paul Sweezy,
https://climateandcapitalism.com/2012/03/10/paul-sweezy-capitalism-versus-the-
environment/

4 - What Happened to the Proletariat, David Rosen,
https://www.counterpunch.org/2015/06/12/whatever-happened-to-the-proletariat/

5 - Wilder Penfield, The Mystery of the Mind, Princeton University Press, p. 31

6 - Quoted in "Scientists say your "mind" isn't confined to your brain, or even your
body", Olivia Goodhill,
https://qz.com/866352/scientists-say-your-mind-isnt-confined-to-your-brain-or-even-
your-body/

7 - Quoted in "You Might Be a Terrorist, Too", Radley Balko,
https://reason.com/2011/02/02/you-might-be-a-terrorist-too/

8 - "Predictive Policing Software Is More Accurate at Predicting Policing Than
Predicting Crime', Ezekiel Edwards,
https://www.aclu.org/blog/criminal-law-reform/reforming-police-practices/predictive-
policing-software-more-accurate

9 - "What's Wrong with Fusion Centers", ACLU, https://www.aclu.org/report/whats-
wrong-fusion-centers-
executive-summary

10 - "Facebook IPO: What the %$#! happened", Julianne Pepitone
https://money.cnn.com/2012/05/23/technology/facebook-ipo-what-went-wrong/
index.htm

11 - "How Amazon Actually Makes Money", The Motley Fool Staff,
https://www.fool.com/investing/2019/02/19/how-amazon-
actually-makes-money.aspx

12 - "Movement Technologists Statement", May First Movement Technology
https://outreach.mayfirst.org/techstatement

13 - ibid, https://outreach.mayfirst.org/techstatement

14 - "Narrative Power Analysis" Patrick Reinsborough and Doyle Canning, https://beautifultrouble.org/theory/narrative-power-analysis/0.79